SEMA SIMAI Springer Series

ICIAM 2019 SEMA SIMAI Springer Series

Volume 2

This sub-series of the SEMA SIMAI Springer Series aims to publish some of the most relevant results presented at the ICIAM 2019 conference held in Valencia in July 2019.

The sub-series is managed by an independent Editorial Board, and will include peer-reviewed content only, including the Invited Speakers volume as well as books resulting from mini-symposia and collateral workshops.

The series is aimed at providing useful reference material to academic and researchers at an international level.

More information about this subseries at http://www.springer.com/series/16499

Grigory Sklyar • Alexander Zuyev
Editors

Stabilization of Distributed Parameter Systems: Design Methods and Applications

Springer

Editors
Grigory Sklyar
Institute of Mathematics
University of Szczecin
Szczecin, Poland

Alexander Zuyev
Max Planck Institute for Dynamics
of Complex Technical Systems
Magdeburg, Germany

Institute of Applied Mathematics
and Mechanics
National Academy of Sciences of Ukraine
Sloviansk, Ukraine

ISSN 2199-3041 ISSN 2199-305X (electronic)
SEMA SIMAI Springer Series
ISSN 2662-7183 ISSN 2662-7191 (electronic)
ICIAM 2019 SEMA SIMAI Springer Series
ISBN 978-3-030-61744-8 ISBN 978-3-030-61742-4 (eBook)
https://doi.org/10.1007/978-3-030-61742-4

This Springer imprint is published by the registered company Springer Nature Switzerland AG.
The registered company address is: Gewerbestrasse 11, 6330 Cham, Switzerland

Preface

This book resulted from presentations and scientific discussions at the Thematic Minisymposium "Stabilization of Distributed Parameter Systems: Design Methods and Applications" within the program of the 9th International Congress on Industrial and Applied Mathematics (ICIAM 2019), held in Valencia from 15 to 19 July 2019.

The contributions collected in this book aim at bringing together recent results on stabilizing control design for different classes of infinite-dimensional dynamical systems described by partial differential equations, functional-differential equations, delay equations, and dynamical systems in abstract spaces. The presentations concern various types of stability notions, such as strong, uniform, exponential, and polynomial stability phenomena. This includes new results in the theory of linear and nonlinear operator semigroups, port-Hamiltonian systems, non-Fourier moment problem, turnpike analysis, and further developments of Lyapunov's direct method.

The scope of the book also covers applications of these theoretical methods to flexible mechanical structures and mathematical models in chemical engineering. Special emphasis is given to the stability analysis of particulate and crystallization processes governed by controlled hyperbolic systems. The presentation is accompanied with the results of numerical simulations.

The eight chapters gathered in this volume present a selection of contributions reported at ICIAM 2019. Every manuscript has undergone a careful peer review process. We express our sincere gratitude to the authors and reviewers for their valuable work. We would also like to thank the members of the editorial team at Springer for their support in the preparation of this volume.

Szczecin, Poland
Magdeburg, Germany
April 2020

Grigory Sklyar
Alexander Zuyev

Contents

About the Editors

Alexander Zuyev received his Ph.D. degree in 2000 from the Institute of Applied Mathematics and Mechanics, National Academy of Sciences of Ukraine (IAMM NASU). He was a visiting scientist at the Abdus Salam International Centre for Theoretical Physics under the aegis of UNESCO and IAEA in Trieste and received the Alexander von Humboldt Research Fellowship at the TU Ilmenau and the University of Stuttgart. Since his habilitation in 2008, he has been working as a Professor at the Donetsk National University and a Leading Researcher and Department Head at IAMM NASU. He is currently with the Max Planck Institute for Dynamics of Complex Technical Systems and the Otto von Guericke University Magdeburg in Germany. Professor Zuyev has authored 2 books and more than 50 articles in the field of mathematical control theory, stability theory, and mathematical problems of mechanics and chemical engineering.

Grigory Sklyar received his Ph.D. degree in 1983 from the Kharkov State University (USSR) and habilitation in 1991 from the B.Verkin Institute for Low Temperature Physics and Engineering, National Academy of Science of Ukraine. He worked as a Professor at the V. N. Karazin Kharkiv National University (1992–1999). Since 1999, he is a Full Professor at the Institute of Mathematics of University of Szczecin (Poland). He also has been working on visiting positions at Technical University of Darmstadt, Institut de Recherche en Communications et Cybernétique de Nantes, Institute for Advanced Study in Mathematics, Hanoi, and others. Professor Sklyar is an author of more than 100 works in the field of Functional Analysis, Differential Equations, and Mathematical Control Theory.

Conditions of Exact Null Controllability and the Problem of Complete Stabilizability for Time-Delay Systems

Pavel Barkhayev, Rabah Rabah, and Grigory Sklyar

Abstract For a class of linear time-delay control systems satisfying the property of completability of the generalized eigenvectors, we prove that the problems of complete stabilizability and exact null controllability are equivalent.

1 Introduction

The problems of controllability and stabilizability are among the central and most investigated in the mathematical control theory; these problems are well studied in many cases. The relation between the problems and these notions themselves depends essentially on specific settings. For example, in finite-dimensional settings, most of the main controllability notions (exact, null, and approximate) are equivalent and imply stabilizability. Moreover, complete stabilizability (stabilizability with an arbitrary decay rate) is equivalent to controllability. In infinite-dimensional settings, the situation is much more sophisticated. First of all, different notions of controllability are not equivalent in general. Second, exact null controllability implies complete stabilizability, but the inverse does not hold in general. In some special cases, if $\{e^{\mathcal{A}t}\}$ is a group [30] or the operators $e^{\mathcal{A}t}$ are surjective for all $t \geq 0$ (see [17, 31]), complete stabilizability implies exact controllability. Finally, exact

P. Barkhayev (✉)
Department of Mathematical Sciences, Norwegian University of Science and Technology, Trondheim, Norway

B.Verkin Institute for Low Temperature Physics and Engineering, Academy of Sciences of Ukraine, Kharkiv, Ukraine
e-mail: pavloba@ntnu.no; barkhayev@ilt.kharkov.ua

R. Rabah
IMT Atlantique, Mines-Nantes, Nantes, France

G. Sklyar
Institute of Mathematics, University of Szczecin, Szczecin, Poland
e-mail: sklar@univ.szczecin.pl

G. Sklyar, A. Zuyev (eds.), *Stabilization of Distributed Parameter Systems: Design Methods and Applications*, SEMA SIMAI Springer Series 2,
https://doi.org/10.1007/978-3-030-61742-4_1

null controllability implies complete stabilizability with bounded feedback (see [23] for the proof and counterexamples).

The problem of exponential stabilizability for linear time-delay systems was considered by many authors, see e.g. [7, 13, 14, 16, 24] and references therein. For the problem of asymptotic non-exponential stabilizability, which appears only for neutral type systems, we refer to [20, 21]. An analysis of relations between exact controllability and exponential stabilizability may be found, for example, in [6, 8, 14, 26].

A rather general class of linear control time-delay systems with distributed delays is described by equation

$$\dot{z}(t) - A_{-1}\dot{z}(t-1) = Lz_t(\cdot) + bu(t), \quad t \geq 0; \tag{1}$$

here, $z(t) \in \mathbb{R}^n$ is state, and we use the notation $z_t(\theta) = z(t+\theta)$, $\theta \in [-1, 0]$, $t \geq 0$; $u(t) \in \mathbb{R}$ is control; the operator L, acting from the Sobolev space $W^{1,2}((-1,0); \mathbb{R}^n)$ to $\mathbb{R}^n$ is bounded, i.e.

$$Lz_t(\cdot) = \int_{-1}^{0} [A_2(\theta)\dot{z}(t+\theta) + A_3(\theta)z(t+\theta)]\,\mathrm{d}\theta,$$

the matrices A_{-1}, A_2, A_3, and b are of appropriate dimensions, and the elements of A_2 and A_3 take values in $L^2(-1,0)$.

For the system (1) of neutral type, the problem of exact controllability is deeply investigated in [18, 22]. It was shown that controllability set coincides with the domain of the state operator corresponding to the system (1) and that exact controllability is equivalent to two rank conditions

$$\operatorname{rank}\big(\Delta(\lambda);\ b\big) = n \quad \text{for all} \quad \lambda \in \mathbb{C}, \tag{2}$$

$$\operatorname{rank}\big(-\mu I + A_{-1};\ b\big) = n \quad \text{for all} \quad \mu \in \mathbb{C}, \tag{3}$$

where $\Delta(\lambda)$ is the characteristic matrix of the system (1):

$$\Delta(\lambda) = -\lambda I + \lambda e^{-\lambda} A_{-1} + \int_{-1}^{0} \big[\lambda e^{\lambda s} A_2(s) + e^{\lambda s} A_3(s)\big]\,\mathrm{d}s. \tag{4}$$

We note that the condition (3) assures the existence of a feedback change of the form $u(t) = v(t) + p_{-1}\dot{z}(t-1)$ such that the matrix $A_{-1} + bp_{-1}$ is non-singular. This means that exact null controllability is equivalent to exact controllability (i.e. the state operator $\mathcal{A}$ defined below by (8)–(9) is the generator of a C_0-group).

Later, in [23, Theorem 6], it was shown that complete stabilizability is equivalent to the condition (2) together with

$$\operatorname{rank}\big(-\mu I + A_{-1};\ b\big) = n \quad \text{for all} \quad \mu \in \mathbb{C}\backslash\{0\}. \tag{5}$$

This means, in particular, that exact controllability implies complete stabilizability for the systems (1) and that if the matrix A_{-1} is non-singular or if rank $\left(A_{-1};\ b\right) = n$, then exact controllability is equivalent to complete stabilizability. However, in the case rank $\left(A_{-1};\ b\right) < n$, the situation is unclear. In [23], we posed a conjecture that complete stabilizability is equivalent to exact null controllability in the general case.

In the special case of retarded systems ($A_{-1} = 0$), complete stabilizability is equivalent to (2), which is called spectral controllability condition. It is well known that exact null controllability implies spectral controllability (see e.g. [25]); however, the inverse has not been proved for general systems. One of the first results on this issue was obtained in [9] by Jacobs and Langenhop, where the conjecture is proved in the case $n = 2$ and $Lf = A_1 f(-h) + A_0 f(0)$, $A_i \in \mathbb{R}^{2\times 2}$. Later, in [12], Marchenko claimed the conjecture for control systems with finitely many delays; however, in [25], it is noticed that his arguments seem to be incomplete, see also [28]. In 1984, Colonius [4] has showed the conjecture in the case $Lf = A_1 f(-h) + A_0 f(0)$ and arbitrary n. His proof is based on the fact that spectrum controllability is equivalent to solvability of the finite spectrum assignment problem. Later, Olbrot and Pandolfi [15] have given an explicit algebraic algorithm of computing a control function which steers any given initial function to zero in finite time for a quite wide class of retarded control systems.

We also note that complete stabilizability of neutral systems, which is exponential stabilizability with arbitrary decay rate, requires that stabilizing feedback operators contain a derivative of the delayed state (as it was noted early, cf. [14, 16]). However, such a class of feedbacks is of theoretical value mostly, since its practical implementation may lead to essential difficulties. To achieve complete stabilizability in case of retarded systems, it is enough to use the feedbacks containing delayed state terms only. It is also worth to mention that the latter class of feedbacks is used to solve the strong stabilizability problem for neutral type systems (see [20, 21] for details).

In the present paper, we show that the property of completeness (completability) of the set of generalized eigenvectors is crucial for equivalence between exact controllability and complete stabilizability. This property allows to represent the steering conditions of controllability as a vector moment problem and investigate its solvability in an appropriate class of functions. If the system (1) is of neutral type, then the family of exponentials corresponding to the moment problem forms the Riesz basis of its closure which gives a powerful tool for investigation [1]. This method was used in [18, 22], where an exhaustive analysis of controllability was given. In the case when system (1) is not neutral (in particular, retarded), the family of exponentials does not possess the Riesz basis property. However, as it was shown in [10] for the case of systems with point-wise delays, some similar conditions of controllability can be obtained. We use the moment problem approach and show that trigonometric moment problems corresponding to systems of the form (1) are solvable on intervals of appropriate length subject to some rank conditions. To show solvability, we construct explicitly systems biorthogonal to exponentials [1, 11].

For a class of completable systems, this method allows us to prove the equivalence of exact null controllability and complete stabilizability and, thus, to show the conjecture posed in [23].

The paper is organized as follows. In Sect. 2, we introduce some notations and definitions and rewrite equivalently the system (1) in the infinite-dimensional model space. In Sect. 3, we introduce the moment problem corresponding to the controllability conditions. We give the conditions of completeness and completability of the generalized eigenvectors. In Sect. 4, we prove the main result on controllability which implies equivalence of complete stabilizability and exact null controllability for system satisfying the completability condition. Besides, we give examples.

2 Preliminaries

Let us consider an initial condition of the form

$$\begin{cases} z(0) = y, \\ z(t) = z_0(t), \quad t \in [-1, 0). \end{cases} \tag{6}$$

In the case $A_{-1} = 0$ for any $y \in \mathbb{R}^n$, $z_0(t) \in L^2((-1,0);\mathbb{R}^n)$ and any control $u(t) \in L^2_{\text{loc}}[0,+\infty)$, there exists a unique solution $z(t)$, $t \geq 0$ of the initial-value problem (1) and (6), which is continuous (see [5]). In the case of neutral type systems ($A_{-1} \neq 0$), the existence of the strong solution is guaranteed for smooth initial states only: $z_0(t) \in W^{1,2}((-1,0);\mathbb{R}^n)$ (see [3], where the neutral operator of a more general form was considered).

These facts naturally lead to consideration of the problem (1) and (6) in the product space

$$M_2(-1,0;\mathbb{R}^n) \stackrel{\text{def}}{=} \mathbb{R}^n \times L^2(-1,0;\mathbb{R}^n),$$

further noted shortly as M_2. Thus, the initial-value problem (1)–(6) may be rewritten as

$$\begin{cases} \dot{x}(t) = \mathcal{A}x(t) + \mathcal{B}u(t), \ t \geq 0, \\ x(0) = x_0; \end{cases} \tag{7}$$

here,

$$\mathcal{A}x(t) = \begin{pmatrix} Lz_t(\cdot) \\ \frac{\mathrm{d}z_t(\theta)}{\mathrm{d}\theta} \end{pmatrix}, \quad x(t) = \begin{pmatrix} y(t) \\ z_t(\cdot) \end{pmatrix}, \tag{8}$$

with

$$\mathcal{D}(\mathcal{A}) = \left\{ \begin{pmatrix} y \\ z_0(\tau) \end{pmatrix} : z_0(\tau) \in W^{1,2}((-1,0);\mathbb{R}^n),\ y = z_0(0) - A_{-1}z_0(-1) \right\}, \tag{9}$$

and $\mathcal{B}u(t) = \begin{pmatrix} bu(t) \\ 0 \end{pmatrix} \in M_2$, $x_0 = \begin{pmatrix} y \\ z_0(\tau) \end{pmatrix}$.

Definition 1 Time-delay system (1) (or the infinite-dimensional system (7)) is said to be exponentially stabilizable if there is a linear feedback operator $\mathcal{F}$:

$$u(t) = \mathcal{F}x(t) = F_{-1}\dot{z}(t-1) + \int_{-1}^{0} [F_2(\theta)\dot{z}(t+\theta) + F_3(\theta)z(t+\theta)]\,d\theta,$$

such that the semigroup $e^{(\mathcal{A}+\mathcal{B}\mathcal{F})t}$ is exponentially stable, i.e. there is a $\omega > 0$ such that

$$\left\| e^{(\mathcal{A}+\mathcal{B}\mathcal{F})t} \right\| \le M_\omega e^{-\omega t}, \quad M_\omega \ge 1. \tag{10}$$

The system (1) is said to be completely stabilizable (or stabilizable with an arbitrary decay rate) if for all $\omega > 0$, there is a linear bounded feedback $\mathcal{F}_\omega$ such that (10) holds.

Definition 2 An initial state $x_0 = (y, z_0(\tau)) \in M_2$ is said to be null controllable by means of the system (1) at time T if there exists a control $u(t) \in L^2[0,T]$ such that $z(t) = z(t, y, z_0, u(t)) \equiv 0$ for $t \in [T-1, T]$.

Since the evolution of the Cauchy problem (7) is described by

$$x(t) = e^{\mathcal{A}t}x_0 + \int_0^t e^{\mathcal{A}(t-\tau)}\mathcal{B}u(\tau)d\tau,$$

the null controllable states satisfy the relation

$$e^{\mathcal{A}T}x_0 = -\int_0^T e^{\mathcal{A}\tau}\mathcal{B}u(T-\tau)\,d\tau, \tag{11}$$

which naturally leads to the notion of attainable set at time T:

$$\mathcal{R}_T = \left\{ \int_0^T e^{\mathcal{A}\tau}\mathcal{B}u(\tau)\,d\tau : u(t) \in L^2[0,T] \right\} \subset M_2. \tag{12}$$

Definition 3 System (1) is said to be exactly null controllable from $F \subset M_2$ at time T if $e^{\mathcal{A}T}F \subset \mathcal{R}_T$.

Definition 4 System (1) is said to be exactly null controllable at time T if it is controllable from M_2, i.e. $\mathrm{Im}(e^{\mathcal{A}T}) \subset \mathcal{R}_T$.

Not every system of the form (1) can be exactly null controllable. For example, if $\det A_{-1} \neq 0$, then $F = \mathcal{D}(\mathcal{A})$ is the maximal possible set of null controllability (see [18, 22] for more details). However, retarded systems ($A_{-1} = 0$) can be exactly null controllable from M_2.

3 The Moment Problem and Completability Property

The eigenvalues $\lambda \in \sigma(\mathcal{A})$ are the zeros of the characteristic function $\det \Delta(\lambda)$, where $\Delta(\lambda)$ is given by (4). Let $\lambda \in \sigma(\mathcal{A})$, the corresponding eigenvector φ_λ of the operator $\mathcal{A}$ is of the form

$$\varphi_\lambda = \begin{pmatrix} (I - e^{\lambda} A_{-1}) x_\lambda \\ e^{\lambda \tau} x_\lambda \end{pmatrix}, \quad x_\lambda \in \mathrm{Ker}\Delta(\lambda),$$

and the eigenvector ψ_λ of the operator $\mathcal{A}^*$ corresponding to $\overline{\lambda}$ is of the form

$$\psi_\lambda = \begin{pmatrix} y_\lambda \\ \left[\overline{\lambda} e^{-\overline{\lambda}\tau} I - A_2^*(\tau) + \int_0^\tau e^{\overline{\lambda}(s-\tau)} \left(A_3^*(s) + \overline{\lambda} A_2^*(s)\right) \mathrm{d}s\right] y_\lambda \end{pmatrix}, \tag{13}$$

where $y_\lambda \in \mathrm{Ker}\Delta^*(\lambda)$.

In (11), we make the change of time $v(\tau) = -u(T - \tau)$ and multiply the relation by ψ_λ:

$$\left\langle e^{\mathcal{A}T} x_0, \psi_\lambda \right\rangle_{M_2} = \int_0^T \left\langle e^{\mathcal{A}\tau} \mathcal{B} v(\tau), \psi_\lambda \right\rangle_{M_2} \mathrm{d}\tau,$$

which gives

$$\left\langle x_0, e^{\overline{\lambda} T} \psi_\lambda \right\rangle_{M_2} = \int_0^T \left\langle \mathcal{B} v(\tau), e^{\overline{\lambda}\tau} \psi_\lambda \right\rangle_{M_2} \mathrm{d}\tau,$$

or

$$e^{\lambda T} \langle x_0, \psi_\lambda \rangle_{M_2} = \langle b, y_\lambda \rangle_{\mathbb{R}^n} \int_0^T e^{\lambda\tau} v(\tau) \mathrm{d}\tau.$$

If the spectral controllability condition (2) holds, then $\langle b, y_\lambda\rangle_{\mathbb{R}^n} \neq 0$, and we obtain the moment problem

$$s_\lambda = \int_0^T e^{\lambda\tau} v(\tau) d\tau, \quad \lambda \in \sigma(\mathcal{A}), \tag{14}$$

where

$$s_\lambda \equiv e^{\lambda T} \langle x_0, \psi_\lambda\rangle_{M_2} (\langle b, y_\lambda\rangle_{\mathbb{R}^n})^{-1}. \tag{15}$$

Remark 1 If $u \in L^2(0, T)$ steers $x_0 \in M_2$ to null at time T, then the corresponding function $v(\cdot)$ solves the moment problem (14). The inverse assertions hold only if the system $\{\psi_\lambda\}_{\lambda\in\sigma(\mathcal{A})}$ is complete in M_2.

Indeed, the sequence $\{s_\lambda\}$ defines uniquely the initial point x_0 by (15), since if $s_\lambda = e^{\lambda T} \langle x_0, \psi_\lambda\rangle_{M_2} (\langle b, y_\lambda\rangle_{\mathbb{R}^n})^{-1} = e^{\lambda T} \langle x_1, \psi_\lambda\rangle_{M_2} (\langle b, y_\lambda\rangle_{\mathbb{R}^n})^{-1}$, then $\langle x_0 - x_1, \psi_\lambda\rangle_{M_2} = 0$, $\lambda \in \sigma(\mathcal{A})$, which gives $x_1 = x_0$. Furthermore, if $x_0 \notin R_T$, it would mean that $h \equiv e^{\mathcal{A}T} x_0 - \int_0^T e^{\mathcal{A}\tau} \mathcal{B}u(T-\tau)d\tau \neq 0$. However, $\langle h, \psi_\lambda\rangle_{M_2} = 0$ due to (14), and thus $h = 0$.

Furthermore, we give conditions of completeness. Let m_λ be the length of the maximal chain of the generalized eigenvectors corresponding to $\lambda \in \sigma(\mathcal{A})$. The generalized eigenspace corresponding to λ is

$$V_\lambda(\mathcal{A}) = \mathrm{Ker}(\mathcal{A} - \lambda I)^{m_\lambda},$$

and the linear span

$$V(\mathcal{A}) = \mathrm{Lin}\{V_\lambda(\mathcal{A}) : \lambda \in \sigma(\mathcal{A})\}$$

is called the generalized eigenspace of $\mathcal{A}$.

The entire function $f(\lambda) = \det \Delta(\lambda)$ is of exponential type and satisfies the condition

$$\frac{\log |f(x_0 + iy)|}{1 + y^2} \in L^1(\mathbb{R}),$$

for some fixed x_0 satisfying $x_0 > \mathrm{Re}\lambda$, for any $\lambda \in \sigma(\mathcal{A})$. This means that $\det \Delta(\lambda)$ belongs to the class C (see e.g. [11, Lecture 16] for more details). It is known that indicator function of $f \in C$ is of the form

$$h_f(\theta) = \lim_{r\to\infty} \frac{\log |f(re^{i\theta})|}{r} = \begin{cases} \alpha_-(f)|\cos\theta|, & \frac{\pi}{2} \le \theta \le \frac{3\pi}{2}, \\ \alpha_+(f)\cos\theta, & \frac{3\pi}{2} \le \theta \le \frac{5\pi}{2}, \end{cases}$$

where $\alpha_-(f)$ and $\alpha_+(f)$ are some constants, and the limit exists almost for all θ.

In [27], a criterion of completeness of generalized eigenvectors is established (in more general settings).

Theorem 1 ([27, Theorem 4.2]) *The generalized eigenvectors of $\mathcal{A}$ are complete in M_2, i.e. $\overline{V(\mathcal{A})} = M_2$ if and only if*

$$\alpha_-(\det \Delta) = n.$$

Let us consider the special case of the operator L:

$$Lf = A_1 f(-1) + \int_{-1}^{0} A_2(\theta) f'(\theta)\, \mathrm{d}\theta + \int_{-1}^{0} A_3(\theta) f(\theta)\, \mathrm{d}\theta, \tag{16}$$

assuming that

$$\operatorname{supp} A_i(\theta) \subset [\alpha, 0], \quad \alpha > -1, i = 2, 3. \tag{17}$$

For this case, Theorem 1 may be reformulated as

Corollary 1 *The generalized eigenvectors of the operator $\mathcal{A}$ of the system (1) with the operator L defined by (16) are complete if and only if the matrix pencil $A_1 + \lambda A_{-1}$ is non-singular; that is, there exists $\lambda_0 \in \mathbb{C}^n$ such that*

$$\det(A_1 + \lambda_0 A_{-1}) \neq 0.$$

Proof We rewrite $\det \Delta(\lambda)$:

$$\det \Delta(\lambda) = f_1(\lambda) + f_2(\lambda), \quad f_1(\lambda) = \det(A_1 + \lambda A_{-1}) \mathrm{e}^{-nh}.$$

Due to the assumption (17): $\alpha_-(f_2) < n$ and since $\alpha_-(\mathrm{e}^{-nh}) = n$, we can conclude that $\alpha_-(\det \Delta) = n$ if and only if $\det(A_1 + \lambda A_{-1}) \not\equiv 0$. □

It is worth to mention that this result may be obtained using the technique similar to [2, Theorem 2].

Definition 5 Control system (1) is said to be completable if there exists a feedback

$$u(t) = \mathcal{P} x(t) = p_{-1} \dot{z}(t-1) + P z_t, \quad p_{-1} \in \mathbb{R}^n, P \in \mathfrak{B}(W^{1,2}((-1,0), \mathbb{R}^n), \mathbb{R}),$$

such that the operator $\mathcal{A} + \mathcal{B}\mathcal{P}$ of the closed-loop system possesses a complete set of generalized eigenvectors.

From Definition 5 and Corollary 1, we obtain the following corollary.

Corollary 2 *The system (1) with the operator L defined by (16) is completable if and only if there exists a feedback of the form*

$$u(t) = p_{-1} \dot{z}(t-1) + p_1 z(t-1), \quad p_{-1}, p_1 \in \mathbb{R}^n,$$

such that the matrix pencil

$$(A_1 + bp_1) + \lambda(A_{-1} + bp_{-1})$$

is non-singular.

4 The Main Result

Consider the special case of the system (1) when $A_{-1} = 0$ and the operator L is defined by (16), that is,

$$\dot{z}(t) = A_1 z(t-1) + \int_{-1}^{0} [A_2(\theta)\dot{z}(t+\theta) + A_3(\theta)z(t+\theta)]\,\mathrm{d}t + bu(t), \tag{18}$$

with

$$\operatorname{supp} A_i(\theta) \subset [\alpha, 0], \quad \alpha > -1, \quad i = 2, 3. \tag{19}$$

We assume that the system (18) is completable, which means, due to Corollary 2, that the pair (A_1, b) is controllable:

$$\operatorname{rank}(-\lambda I + A_1; b) = n, \quad \lambda \in \mathbb{C}. \tag{20}$$

Theorem 2 *The system (18) satisfying the rank condition (20) is exactly null controllable from any initial state $x_0 \in M_2$ if and only if it is spectrally controllable.*

Proof We divide the proof of the proposition into several steps.

Step 1 We fix n distinct nonzero numbers $\{a_i\}$ and apply a change of feedback $u(t) = v(t) + \hat{A}_1 z(t-1)$ and a change of coordinates such that the system takes the form (18) with

$$A_1 = \operatorname{diag}(a_i), \quad b = (1, \ldots, 1)^*.$$

Besides, we assume, without loss of generality, that the eigenvalues of the corresponding operator $\mathcal{A}$ are simple. Indeed, only a finite number of eigenvalues may be multiple, and thus there exists a feedback change of the form

$$u(t) = v(t) + \int_{-1}^{0} \left[\hat{A}_2(\theta)\dot{z}(t+\theta) + \hat{A}_3(\theta)z(t+\theta)\right]\mathrm{d}t,$$

which makes the spectrum of $\mathcal{A}$ simple.

Step 2 The characteristic function of the system $\dot{z}(t) = A_1 z(t-1)$ has the form $\det(-\lambda I + e^{-\lambda} A_1) = \prod_{j=1}^{n}(-\lambda + a_j e^{-\lambda})$. Each of the entire functions $g_j(\lambda) = -\lambda + a_j e^{-\lambda}$ possesses infinitely many zeros which we denote as $\{\tilde{\lambda}_k^j\}_{k\in\mathbb{Z}}$ assuming that λ_0^j is the only real zero of $g_j(\lambda)$ and $\mathrm{Im}\lambda_k^j < \mathrm{Im}\lambda_s^j$ as $k < s$. We also consider the circles $L_k^j(r_0)$ of fixed radius $r_0 = \frac{1}{3}\min\{|\tilde{\lambda}_{k_1}^{j_1} - \tilde{\lambda}_{k_2}^{j_2}|, (k_1, j_1) \neq (k_2, j_2)\}$ centered at $\tilde{\lambda}_k^j$.

As in [19, Theorem 2], it can be shown that there exists $N \in \mathbb{N}$ such that for any j, satisfying $|j| \geq N$, the characteristic function $\det \Delta(\lambda)$ of (18) possesses a zero in the circle $L_k^j(r_0)$.

We denote the zeros of the characteristic function $\det \Delta(\lambda)$ as $\{\lambda_k^j\}$. The eigenvectors $\{\psi_k^j\}$ of the operator $\mathcal{A}^*$ are given by (13). Due to the spectral controllability condition, we have $\left\langle b, y_k^j \right\rangle_{\mathbb{R}^n} \neq 0$, and thus we can normalize ψ_k^j such that $\left\langle b, y_k^j \right\rangle_{\mathbb{R}^n} = 1$. Then, the moment problem (14) for Eq. (18) takes the form

$$s_k^j = \int_0^T e^{\lambda_k^j \tau} u(\tau) d\tau, \quad j = \overline{1,n}, k \in \mathbb{Z}, \tag{21}$$

where

$$s_k^j \equiv e^{\lambda_k T} \left\langle x_0, \psi_k^j \right\rangle_{M_2}.$$

The system $\{\psi_k^j\}$ is complete in M_2 due to Corollary 1, and thus solvability of (21) is equivalent to exact null controllability of the initial state $x_0 \in M_2$. Furthermore, we show that the moment problem (21) is solvable in class $L^2[0, T]$ for $T > T_0$.

Step 3 Consider the entire function $g(\lambda) = -\lambda + a e^{-\lambda}$, $\lambda \in \mathbb{C}$. We denote its zeros as $\{\lambda_k\}_{k\in\mathbb{Z}}$ assuming that λ_0 is the only real zero of $g(\lambda)$ and $\mathrm{Im}\lambda_k < \mathrm{Im}\lambda_s$ as $k < s$. Consider the family of exponentials

$$\{e^{\lambda_k t}, \quad t \in [0, T], \ \lambda_k : \lambda_k e^{\lambda_k} = 1\}. \tag{22}$$

There exists $T_0 > 0$ such that the family (22) is minimal in $L^2[0, T]$ for any $T > T_0$ (see e.g. [29]). Thus, there exists a biorthogonal family to (22) in $L^2[0, T]$, and we construct it explicitly.

First, we define $\widetilde{f_s}(t) = e^{-\overline{\lambda_s} t} - e^{-\overline{\lambda_{s+1}} t}$, $s \in \mathbb{Z}$, and

$$\widetilde{f_s^1}(t) = \begin{cases} \widetilde{f_s}(t), & t \in [0, 1], \\ 0, & t \in (1, T], \end{cases} \qquad \widetilde{f_s^2}(t) = \begin{cases} 0, & t \in [0, T-1), \\ \widetilde{f_s}(t - T + 1), & t \in [T-1, T]. \end{cases} \tag{23}$$

Furthermore, we construct in the space $L^2[0,T]$ $(T > 1)$ the biorthogonal to $\{e^{\lambda_k t}\}_{k\in\mathbb{Z}}$ family $\{f_s(t)\}_{s\in\mathbb{Z}}$:

$$f_s(t) = \beta_s^1 \widetilde{f}_s^1(t) + \beta_s^2 \widetilde{f}_s^2(t), \quad \beta_s^i \in \mathbb{C}, \tag{24}$$

where β_s^i are obtained from the condition $\langle e^{\lambda_k t}, f_s(t)\rangle_{L^2[0,T]} = \delta_{ks}$:

$$\beta_s^1 = -\frac{\widetilde{\lambda}_s}{(\widetilde{\lambda}_{s+1} - \widetilde{\lambda}_s)\alpha_s}, \qquad \beta_s^2 = \frac{\widetilde{\lambda}_s \widetilde{\lambda}_{s+1}}{(\widetilde{\lambda}_{s+1} - \widetilde{\lambda}_s)\alpha_s}, \tag{25}$$

where $\widetilde{\lambda}_s = (\lambda_s)^{T-1}$, $\alpha_s = 1 + \frac{1}{\lambda_s}$.

We estimate the growth of the constructed family $\{f_s(t)\}$:

$$\|f_s(t)\|_{L^2[0,T]} \le |\beta_s^1| \|\widetilde{f}_s^1(t)\|_{L^2[0,T]} + |\beta_s^2| \|\widetilde{f}_s^2(t)\|_{L^2[0,T]} = \left(|\beta_s^1| + |\beta_s^2|\right) \|\widetilde{f}_s(t)\|_{L^2[0,1]}.$$

Since $|\alpha_s| \sim 1$, $|\lambda_s - \lambda_{s+1}| \sim 1$, and $|\lambda_s| \sim s$, one has $|\beta_s^1| \sim s^{T-1}$, and $|\beta_s^2| \sim s^{2(T-1)}$. The norm of $\widetilde{f}_s(t)$ may be computed explicitly:

$$\|\widetilde{f}_s(t)\|^2_{L^2[0,1]} == \frac{-\left[(|\lambda_s|^2 - 1)\mathrm{Re}\lambda_{s+1} + (|\lambda_{s+1}|^2 - 1)\mathrm{Re}\lambda_s\right] |\lambda_s - \lambda_{s+1}|^2}{2\mathrm{Re}\lambda_s \mathrm{Re}\lambda_{s+1} |\overline{\lambda}_s + \lambda_{s+1}|^2}.$$

Since $|\mathrm{Re}\lambda_s| \sim \ln s$ and $|\overline{\lambda}_s + \lambda_{s+1}| \sim 1$, one has

$$\|\widetilde{f}_s(t)\|_{L^2[0,1]} \sim s(\ln s)^{-\frac{1}{2}}. \tag{26}$$

Thus, there exist $C > 0$ and $s_0 \in \mathbb{N}$ such that for any $s : |s| > s_0$, one has

$$\|f_s(t)\|_{L^2[0,T]} \le C s^{1+2(T-1)} (\ln s)^{-\frac{1}{2}}. \tag{27}$$

Step 4 We consider the family of exponentials corresponding to eigenvalues $\{\widetilde{\lambda}_k^j\}$, $k \in \mathbb{Z}$, $j = \overline{1,n}$ constructed above:

$$\{e^{\lambda_k^j t} : t \in [0,T],\ \lambda_k^j e^{\lambda_k^j} = a_j,\ j = \overline{1,n},\ k \in \mathbb{Z}\}, \tag{28}$$

and the corresponding moment problem

$$s_k^j = \int_0^T e^{\lambda_k^j \tau} u(\tau) d\tau, \quad j = \overline{1,n},\ k \in \mathbb{Z}. \tag{29}$$

The family (28) is minimal in $L^2[0,T]$ for $T > n$ (see e.g. [29]). Thus, there exists a biorthogonal family, and below we construct and estimate it, which allows us to show solvability of the moment problem (29) in $L^2[0,T]$.

Consider the Hilbert space $E = \prod_{j=1}^{n} L^2[0,T] = \{(x_1,\ldots,x_n) : x_j \in L^2[0,T]\}$ with the norm $\|x\|_E = \sum_{j=1}^n \|x_j\|_{L^2[0,T]}$ and its closed subspace

$$E_1 = \left\{(x_1,\ldots,x_n) : x_j \in \overline{\mathrm{Lin}\{e^{\lambda_k^j t} : k \in \mathbb{Z}\} \subset L^2[0,T]}\right\} \subset E. \tag{30}$$

Consider a linear operator D acting from E_1 to $L^2[0,T]$ as

$$Dx = \sum_{j=1}^{n} x_j. \tag{31}$$

Obviously, the operator D is bounded: $\|Dx\|_{L^2[0,T]} \le \sum_{j=1}^n \|x_j\|_{L^2[0,T]} = \|x\|_{E_1}$. The range of D is

$$E_2 = \overline{\mathrm{Lin}\{e^{\lambda_k^j t} : \ j = \overline{1,n},\ k \in \mathbb{Z}\} \subset L^2[0,T]}. \tag{32}$$

Moreover, since the system of exponentials $\{e^{\lambda_k^j t} : j = \overline{1,n},\ k \in \mathbb{Z}\}$ is minimal in $L^2[0,T]$, D is bijective as an operator from E_1 to E_2, and due to the Banach theorem on the inverse operator, $D^{-1} : E_2 \to E_1$ is bounded.

Let us consider the following two systems of functions:

$$\begin{aligned} \varphi_k^j(t) &= \Big(\underbrace{0,\ldots,0}_{j-1}, e^{\lambda_k^j t}, 0,\ldots,0\Big) \in E_1, \\ \psi_k^j(t) &= \Big(\underbrace{0,\ldots,0}_{j-1}, f_k^j, 0,\ldots,0\Big) \in E, \end{aligned} \tag{33}$$

where for each fixed j, the family $\{f_k^j\}$, constructed by (24), is biorthogonal to $\{e^{\lambda_k^j t}\}$ in $L^2[0,T]$. By construction, we have $\left\langle \varphi_{k_1}^{j_1}, \psi_{k_2}^{j_2} \right\rangle_E = \delta_{(k_1,j_1)}^{(k_2,j_2)}$. We denote by $\widetilde{\psi}_k^j$ the orthogonal projection of ψ_k^j onto subspace E_1, i.e. $\psi_k^j = \widetilde{\psi}_k^j + d_k^j$, where $d_k^j \in (E_1)^\perp$. The family $\{\widetilde{\psi}_k^j\}$ is biorthogonal to $\{\varphi_{k_1}^{j_1}\}$ in E_1:

$$\left\langle \varphi_{k_1}^{j_1}, \widetilde{\psi}_{k_2}^{j_2} \right\rangle_{E_1} = \left\langle \varphi_{k_1}^{j_1}, \psi_{k_2}^{j_2} \right\rangle_E = \delta_{(k_1,j_1)}^{(k_2,j_2)}; \tag{34}$$

moreover, it is well known that $\left\| \widetilde{\psi}_k^j \right\|_{E_1} \le \left\| \psi_k^j \right\|_E$.

The family $\{g_k^j(t)\} = \{(D^{-1})^* \widetilde{\psi}_k^j\}$ is biorthogonal to $\{e^{\lambda_k^j t}\}$ in $E_2 \subset L^2[0, T]$:

$$\left\langle e^{\lambda_{k_1}^{j_1} t}, (D^{-1})^* \widetilde{\psi}_{k_2}^{j_2} \right\rangle_{E_2} = \left\langle D^{-1} e^{\lambda_{k_1}^{j_1} t}, \widetilde{\psi}_{k_2}^{j_2} \right\rangle_{E_1} = \left\langle \varphi_{k_1}^{j_1}, \widetilde{\psi}_{k_2}^{j_2} \right\rangle_{E_1} = \delta_{(k_1, j_1)}^{(k_2, j_2)}. \tag{35}$$

Since D^{-1} is bounded, the adjoint operator $(D^{-1})^*$ is bounded as well, and $\|(D^{-1})^*\| = \|D^{-1}\|$, and thus, we can estimate

$$\left\| g_k^j(t) \right\|_{L^2[0,T]} \leq \left\| (D^{-1})^* \right\| \left\| \widetilde{\psi}_k^j \right\|_{E_1} \leq C \left\| \psi_k^j \right\|_E = C \left\| f_k^j \right\|_{L^2[0,T]}. \tag{36}$$

Step 5 The sequences $\{\lambda_k^j\}$ and $\{\tilde{\lambda}_k^j\}$ are such that $|\lambda_k^j - \tilde{\lambda}_k^j| \to 0$ as $k \to \infty$ for any $j \in \overline{1, n}$. Let $T > 0$ be such that the system $\{e^{\tilde{\lambda}_k^j t}\}$ is minimal in $L^2(0, T)$ and $\{g_k^j(t)\}$ is the biorthogonal system constructed above. Then, there exists a system $\{h_k^j(t)\} \subset L^2(0, T)$ biorthogonal to $\{e^{\lambda_k^j t}\}$ such that

$$\|g_k^j(t) - h_k^j(t)\|_{L^2(0,T)} \to 0, \quad k \to \infty.$$

Due to the estimate on $\{s_k\}$, the series

$$u(t) = \sum_k s_k^j \tilde{h}_k^j(t) \tag{37}$$

converges and $u(t)$ solves the problem (21). □

Corollary 3 *The system (18) satisfying the rank condition (20) is completely stabilizable if and only if it is exactly null controllable.*

Remark 2 The completability condition (20) is not necessary for exact null controllability.

Indeed, let us consider the system

$$\dot{z}(t) = \begin{pmatrix} 0 & 0 \\ 0 & 1 \end{pmatrix} z(t-1) + \begin{pmatrix} 0 & 1 \\ 0 & 0 \end{pmatrix} z(t) + \begin{pmatrix} 0 \\ 1 \end{pmatrix} u(t).$$

It is not completable; however, for any initial state, it is possible to construct a control steering it to zero.

Remark 3 Completeness (completability) does not imply spectral controllability.

The system

$$\dot{z}(t) = \begin{pmatrix} 1 & 0 \\ 0 & 1 \end{pmatrix} z(t-1) + \begin{pmatrix} 0 \\ 1 \end{pmatrix} u(t)$$

is complete, and however it is not controllable.

Acknowledgement The first author was supported by the Norwegian Research Council project "COMAN" No. 275113.

References

1. Avdonin, S.A., Ivanov, S.A.: Families of Exponentials: The Method of Moments in Controllability Problems for Distributed Parameter Systems. Cambridge University Press, Cambridge (1995)
2. Bartosiewicz, Z.: Density of images of semigroup operators for linear neutral functional differential equations. J. Differ. Equ. **38(2)**, 161–175 (1980)
3. Burns, J.A., Herdman, T.L., Stech, H.W.: Linear functional-differential equations as semigroups on product spaces. SIAM J. Math. Anal. **14(1)**, 98–116 (1983)
4. Colonius, F.: On approximate and exact null controllability of delay systems. Syst. Control Lett. **5(3)**, 209–211 (1984)
5. Delfour, M.C.: The largest class of hereditary systems defining a C_0 semigroup on the product space. Canadian J. Math. **32**(4), 969–978 (1980)
6. Dusser, X., Rabah, R.: On exponential stabilizability of linear neutral systems. Math. Probl. Eng. **7**(1), 67–86 (2001)
7. Hale, J.K., Verduyn Lunel, S.M.: Introduction to Functional-Differential Equations. Applied Mathematical Sciences, vol. 99. Springer, New York (1993)
8. Ito, K., Tarn, T.J.: A linear quadratic optimal control for neutral systems. Nonlinear Anal. **9**(7), 699–727 (1985)
9. Jacobs, M., Langenhop, C.E.: Criteria for function space controllability of linear neutral systems. SIAM J. Control Optim. **14**(6), 1009–1048 (1976)
10. Khartovskii, V.E., Pavlovskaya, A.T.: Complete controllability and controllability for linear autonomous systems of neutral type. Autom. Remote Control **(5)**, 769–784 (2013)
11. Levin, B.Ya.: Lectures on Entire Functions. Translations of Mathematical Monographs, vol. 150. American Mathematical Society, Providence (1996)
12. Marchenko, V.M.: On the controllability of zero function of time lag systems. Problems Control Inform. Theory/Problemy Upravlen. Teor. Inform. **8**(5–6), 421–432 (1979)
13. Michiels, W., Niculescu, S.-I.: Stability and Stabilization of Time-Delay Systems: An Eigenvalue-Based Approach. Advances in Design and Control, vol. 12. Society for Industrial and Applied Mathematics, Philadelphia (2007)
14. O'Connor, D.A., Tarn, T.J.: On stabilization by state feedback for neutral differential equations. IEEE Trans. Autom. Control **28**(5), 615–618 (1983)
15. Olbrot, A.W., Pandolfi, L.: Null controllability of a class of functional-differential systems. Int. J. Control **47**(1), 193–208 (1988)
16. Pandolfi, L.: Stabilization of neutral functional differential equations. J. Optim. Theory Appl. **20**(2), 191–204 (1976)
17. Rabah, R., Karrakchou, J.: On exact controllability and complete stabilizability for linear systems in Hilbert spaces. Appl. Math. Lett. **10**(1), 35–40 (1997)
18. Rabah, R., Sklyar, G.M.: The analysis of exact controllability of neutral-type systems by the moment problem approach. SIAM J. Control Optim. **46**(6), 2148–2181 (2007)
19. Rabah, R., Sklyar, G.M., Rezounenko, A.V.: Stability analysis of neutral type systems in Hilbert space. J. Differ. Equ. **214**(2), 391–428 (2005)
20. Rabah, R., Sklyar, G.M., Rezounenko, A.V.: On strong regular stabilizability for linear neutral type systems. J. Differ. Equ. **245**(3), 569–593 (2008)
21. Rabah, R., Sklyar, G.M., Barkhayev, P.Yu.: Stability and stabilizability of mixed retarded-neutral type systems. ESAIM Control Optim. Calc. Var. **18**(3), 656–692 (2012)
22. Rabah, R., Sklyar, G.M., Barkhaev, P.Yu.: On the problem of the exact controllability of neutral-type systems with delay. Ukrain. Mat. Zh. **68**(6), 800–815 (2016)

23. Rabah, R., Sklyar, G.M., Barkhaev, P.Yu.: Exact null controllability, complete stabilizability and continuous final observability of neutral type systems. Int. J. Appl. Math. Comput. Sci. **27**(3), 489–499 (2017)
24. Richard, J.-P.: Time-delay systems: an overview of some recent advances and open problems. Autom. J. IFAC **39**(10), 1667–1694 (2003)
25. Salamon, D.: On controllability and observability of time delay systems. IEEE Trans. Autom. Control **29**(5), 432–439 (1984)
26. Salamon, D.: Control and Observation of Neutral Systems. Research Notes in Mathematics, vol. 91. Pitman, Boston (1984)
27. Verduyn Lunel, S.M., Yakubovich, D.V.: A functional model approach to linear neutral functional-differential equations. Integral Equ. Oper. Theory **27**(3), 347–378 (1997)
28. Watanabe, K.: Further study of spectral controllability of systems with multiple commensurate delays in state variables. Int. J. Control **39**(3), 497–505 (1984)
29. Young, R.M.: An Introduction to Nonharmonic Fourier Series. Pure and Applied Mathematics, vol. 93. Academic, New York (1980)
30. Zabczyk, J.: Mathematical Control Theory: An Introduction. Systems & Control: Foundations & Applications, Birkhäuser Boston, Boston (1992)
31. Zeng, Y., Xie, Z., Guo, F.: On exact controllability and complete stabilizability for linear systems. Appl. Math. Lett., **26**(7), 766–768 (2013)

The Finite-Time Turnpike Phenomenon for Optimal Control Problems: Stabilization by Non-smooth Tracking Terms

Martin Gugat, Michael Schuster, and Enrique Zuazua

Abstract In this paper, problems of optimal control are considered where in the objective function, in addition to the control cost, there is a tracking term that measures the distance to a desired stationary state. The tracking term is given by some norm, and therefore it is in general not differentiable. In the optimal control problem, the initial state is prescribed. We assume that the system is either exactly controllable in the classical sense or nodal profile controllable. We show that both for systems that are governed by ordinary differential equations and for infinite-dimensional systems, for example, for boundary control systems governed by the wave equation, under certain assumptions, the optimal system state is steered exactly to the desired state after finite time.

1 Introduction

Since the turnpike phenomenon has been studied by P. A. Samuelson in mathematical economics in 1949 (see [2]), it has been analyzed in various contexts, see, for example, [1], [17], and [18]. For optimal control problems with partial differential equations, it has been studied in [13] and [15], where distributed control is considered for linear quadratic optimal control problems. Problems of optimal

M. Gugat (✉) · M. Schuster
Lehrstuhl für Angewandte Analysis, Department Mathematik, Friedrich-Alexander Universität Erlangen-Nürnberg (FAU), Erlangen, Germany
e-mail: martin.gugat@fau.de; michi.schuster@fau.de

E. Zuazua
Chair in Applied Analysis, Alexander von Humboldt-Professorship, Department of Mathematics, Friedrich-Alexander-Universität Erlangen-Nürnberg, Erlangen, Germany

Chair of Computational Mathematics, Fundación Deusto, University of Deusto, Bilbao, Basque, Spain

Departamento de Matemáticas, Universidad Autónoma de Madrid, Madrid, Spain
e-mail: enrique.zuazua@fau.de

G. Sklyar, A. Zuyev (eds.), *Stabilization of Distributed Parameter Systems: Design Methods and Applications*, SEMA SIMAI Springer Series 2,
https://doi.org/10.1007/978-3-030-61742-4_2

boundary control are studied in [6], [7] and [10]. In [14], both integral- and measure-turnpike properties are considered. The turnpike phenomenon for linear quadratic optimal control problems with time-discrete systems is studied in [4]. In [5], linear quadratic optimal control problems governed by general evolution equations are considered, and exponential sensitivity and turnpike analysis are studied. An overview on the turnpike phenomenon is given in the monograph [19].

In this paper, we consider integral turnpike properties for problems where the system is exactly controllable, and in the objective function, an L^1-norm or L^2-norm tracking term appears. We show that the resulting optimal controls have a finite-time turnpike structure, that is, the optimal state reaches the static desired state (that we also refer to as the *turnpike* and that does not depend on time) exactly in finite time.

These turnpike results are also useful for numerical computations since they show that for sufficiently large time horizons T, sufficiently accurate approximations of the optimal state/control pairs should also be identical to the desired state with the corresponding constant control most of the time.

The finite-time (or exact) turnpike property for continuous-time systems has already been discussed in [3] as an assumption in the context of nonlinear model predictive control for a finite-dimensional system that is governed by an ordinary differential equation. Here, the aim is to prove convergence in model predictive control. As an application, a problem of optimal fish harvesting control is studied.

This paper has the following structure. In order to illustrate the situation, first we consider optimal control problems that are governed by ordinary differential equations. In these problems, the L^1-norm appears in the tracking term in the objective function. We show that if the weight of the tracking term (i.e. the penalty parameter) is sufficiently large, then the optimal states and controls have a finite-time turnpike structure.

In the next section, we present a finite-time turnpike result for optimal control problems with an abstract infinite-dimensional system. First, we consider the case where the system is exactly controllable. We consider an optimal control problem where the tracking term is given by a certain maximum norm. We show that if the weight of the tracking term is sufficiently large, the solution has a finite-time turnpike structure.

Then, we consider the case where the system is nodal profile exactly controllable. We consider an optimal control problem where the tracking term for the nodal profiles is given by an L^2-norm. We show that if the weight of the tracking term is sufficiently large, then the solution has a finite-time turnpike structure for the nodal profiles.

Finally, we return to the case where the system is exactly controllable. We consider an optimal control problem where the tracking term is given by a weighted L^1-norm that has a singularity at $t_0 > 0$. We also show that, in this case, the solution has a finite-time turnpike structure.

In Sect. 4, examples are presented, where the results from the previous section are applicable. Section 5 contains conclusion.

2 Optimal Control Problems with Ordinary Differential Equation

We start with optimal control problems with systems that are governed by ordinary differential equations. We show that for such systems, L^1-tracking terms in the objective function can lead to finite-time turnpike structures.

Example 1 We start with a system similar to the motivating example in [8] that is governed by an ordinary differential equation. Let $\gamma > 0$ be given. For $T > 0$ sufficiently large (this will be specified later), we consider the problem

$$(\mathbf{OC})_T \begin{cases} \min\limits_{u \in L^2(0,T)} \int_0^T \frac{1}{2}|u(t)|^2 + |u(t)| + \gamma\, |y(t)|\, dt \text{ subject to} \\ y(0) = -1,\ y'(t) = y(t) + \exp(t)\, u(t). \end{cases}$$

The corresponding optimal control problem where the initial condition does not appear is

$$(\mathbf{OC})^{(\sigma)} \begin{cases} \min\limits_{u \in L^2(0,T)} \int_0^T \frac{1}{2}|u(t)|^2 + |u(t)| + \gamma\, |y(t)|\, dt \text{ subject to} \\ y'(t) = y(t) + \exp(t)\, u(t). \end{cases}$$

The solution of $(\mathbf{OC})^{(\sigma)}$ (that we call the turnpike) is zero, that is, $y^{(\sigma)} = 0$ and $u^{(\sigma)} = 0$. The results about the solution of $(\mathbf{OC})_T$ are summarized in the following lemma.

Lemma 1 *For $\gamma > 0$, define $t_0 > 0$ as the minimal value where*

$$(t_0 - 1)\exp(t_0) = \frac{1}{\gamma} - 1.$$

Assume that $T > t_0$ and (even)

$$\gamma\, e^T \geq 1 + \gamma\, e^{t_0}. \tag{2.1}$$

Define

$$\hat{u}(t) = \gamma(e^{t_0} - e^t) \geq 0 \text{ for } t \in (0, t_0],\ \hat{u}(t) = 0 \text{ for } t > t_0. \tag{2.2}$$

Then, for the state $\hat{y}$ generated by $\hat{u}$ for $t \geq t_0$, we have $\hat{y}(t) = 0$. Moreover, for all $t \in (0, T)$, we have $\hat{y}(t) \leq 0$.

The control $\hat{u}$ as defined in (2.2) is the unique solution of $(\mathbf{OC})_T$.

Proof Let a control $u \in L^2(0, T)$ be given. Then, for the corresponding state y, we have

$$y(t) = e^t \left[-1 + \int_0^t u(\tau)\, d\tau \right]. \tag{2.3}$$

Note that for the optimal control, we have $y(t) \le 0$. (If $y(t_0) = 0$, we can continue with the zero control.) Moreover, we have $u(t) \ge 0$. (Otherwise, instead of decreasing the state, it is also better to switch off the control). Hence, it suffices to consider the feasible controls $u(t) \ge 0$ that satisfy the moment inequality

$$\int_0^T u(\tau)\, d\tau \le 1. \tag{2.4}$$

Due to the definition of t_0 and (2.2), we have

$$\int_0^T \hat{u}(\tau)\, d\tau = 1. \tag{2.5}$$

Then, for $t \in (0, t_0)$, we have

$$\hat{y}(t) = e^t \left[-1 + \int_0^t u(\tau)\, d\tau \right] = \gamma\, t\, e^{t+t_0} - \gamma\, e^{2t} + (\gamma - 1) e^t \le 0,$$

and for $t \ge t_0$, we have $\hat{y}(t) = 0$.

With the representation (2.3), for all feasible controls $u \ge 0$ where $y \le 0$, integration by parts yields

$$J_{(0,T)}(u,\, y) = \int_0^T \left[\tfrac{1}{2} |u(t)|^2 + u(t) - \gamma\, y(t) \right] dt$$

$$= \int_0^T \left\{ \tfrac{1}{2} |u(t)|^2 + u(t) + \gamma\, e^t \left[1 - \int_0^t u(\tau)\, d\tau \right] \right\} dt$$

$$= \int_0^T u(t)\, dt + \int_0^T \tfrac{1}{2} |u(t)|^2 + \gamma\, e^t \left[1 - \int_0^t u(\tau)\, d\tau \right] |_{t=0}^{T} + \int_0^T \gamma\, e^t\, u(t)\, dt$$

$$= \int_0^T u(t)\,dt - \gamma + \gamma\,\mathrm{e}^T \left[1 - \int_0^T u(\tau)\,d\tau\right] + \int_0^T \left[\tfrac{1}{2}|u(t)|^2 + \gamma\,\mathrm{e}^t\,u(t)\right] dt$$

$$= (\gamma\,\mathrm{e}^T - 1)\left[1 - \int_0^T u(\tau)\,d\tau\right] + \int_0^T \left[\tfrac{1}{2}|u(t)|^2 + \gamma\,\mathrm{e}^t\,u(t)\right] dt + 1 - \gamma.$$

If T is sufficiently large in the sense that (2.1) holds, due to the L^1-norm that appears in the objective function, the solution has an exact turnpike structure where the system is steered to zero in the finite time t_0 that is independent of T and remains there for $t \in (t_0, T)$. This can be seen as follows. Let $u(t) = \hat{u}(t) + \delta(t)$ with $\hat{u}$ as defined in (2.2) and $\int_0^T \delta(\tau)\,d\tau \le 1 - \int_0^T \hat{u}(\tau)\,d\tau = 0$, where the last equation follows from (2.5) and $\delta(t) \ge 0$ for $t \ge t_0$. Due to (2.1), we have

$$J_{(0,T)}(u, y) = (\gamma\,\mathrm{e}^T - 1)\left[-\int_0^T \delta(\tau)\,d\tau\right]$$

$$+ \int_0^T \left[\tfrac{1}{2}|\hat{u}(t) + \delta(t)|^2 + \gamma\,\mathrm{e}^t\,(\hat{u}(t) + \delta(t))\right] dt + 1 - \gamma$$

$$\ge (\gamma\,\mathrm{e}^T - 1)\left[-\int_0^T \delta(\tau)\,d\tau\right] + \int_0^T \tfrac{1}{2}\hat{u}(t)^2 + \gamma\,\mathrm{e}^t\,\hat{u}(t)\,dt + \int_0^T (\hat{u}(t) + \gamma\,\mathrm{e}^t)\,\delta(t)\,dt + 1 - \gamma$$

$$= (\gamma\,\mathrm{e}^T - 1)\left[-\int_0^T \delta(\tau)\,d\tau\right] + J_{(0,T)}(\hat{u}, \hat{y}) + \int_0^{t_0} \gamma\mathrm{e}^{t_0}\,\delta(t)\,dt + \int_{t_0}^T \gamma\,\mathrm{e}^t\,\delta(t)\,dt$$

$$\ge (\gamma\,\mathrm{e}^T - 1)\left[-\int_0^T \delta(\tau)\,d\tau\right] + J_{(0,T)}(\hat{u}, \hat{y}) + \int_0^{t_0} \gamma\,\mathrm{e}^{t_0}\,\delta(t)\,dt + \int_{t_0}^T \gamma\,\mathrm{e}^{t_0}\,\delta(t)\,dt$$

$$= J_{(0,T)}(\hat{u}, \hat{y}) + (\gamma\,\mathrm{e}^T - \gamma\,\mathrm{e}^{t_0} - 1)\left[-\int_0^T \delta(\tau)\,d\tau\right].$$

Since $\gamma\,\mathrm{e}^T - \gamma\,\mathrm{e}^{t_0} - 1 \ge 0$, this implies that $\hat{u}$ as defined in (2.2) is the optimal control. Thus, we have proved Lemma 1. □

Consider the value t_0 as a function of γ, $t_0 = t_0(\gamma)$. Then, we have $t_0(1) = 1$ and

$$\lim_{\gamma \to \infty} t_0(\gamma) = 0.$$

In Example 2, we present numerical approximations for the optimal states and controls for three values of γ.

2.1 A More General Result for Scalar Ordinary Differential Equations

Now, we consider an optimal control problem with the same objective function and a more general ordinary differential equation. In this problem, we also prescribe a terminal condition. At the end of the section, we will present sufficient conditions that imply that if the penalty parameter γ is sufficiently large, the terminal state is reached before the final time.

Let continuous functions f, g from $[0, \infty)$ to the real numbers be given. Assume that for all $t \geq 0$, we have $f(t) > 0$, and $g(t) > 0$. Let $\gamma \geq 1$ and $\alpha < 0$ be given. For a finite time horizon $T > 0$, we consider the problem

$$(\mathbf{OC})_T \begin{cases} \min\limits_{u(t)\in L^2(0,T), y(t)\in AC(0,T)} \int_0^T \frac{1}{2}|u(t)|^2 + |u(t)| + \gamma\, |y(t)|\, dt \text{ subject to} \\ y(0) = \alpha, \;\; y'(t) = f(t)\, y(t) + g(t)\, u(t) \\ y(T) = 0. \end{cases}$$

Here, again the solution of the corresponding optimal control problem without the initial and the terminal conditions (the turnpike) is zero, that is, $y^{(\sigma)} = 0$ and $u^{(\sigma)} = 0$. Note that the turnpike is compatible with the terminal constraint $y(T) = 0$. In the following theorem, we present the optimal control for $(\mathbf{OC})_T$, which has a similar structure as in the previous example.

Theorem 1 *Define*

$$F(t) = \exp\left(\int_0^t f(s)\, ds\right), \;\; H(t) = \int_0^t F(\tau)\, d\tau.$$

We have

$$y(t) = F(t) \left[\alpha + \int_0^t \frac{g(\tau)}{F(\tau)}\, u(\tau)\, d\tau\right]. \tag{2.6}$$

Define

$$\hat{u}(t) = \max\left\{0,\ \left[-1 - \gamma\,\frac{g(t)\,H(t)}{F(t)} + \lambda \frac{g(t)}{F(t)}\right]\right\}, \tag{2.7}$$

where the number $\lambda > 0$ is chosen such that

$$\int_0^T \hat{u}(\tau)\,\frac{g(\tau)}{F(\tau)}\,d\tau = -\alpha. \tag{2.8}$$

Then, the unique optimal control that solves $(\mathbf{OC})_T$ *is equal to* $\hat{u}(t)$.

Proof Since $y(0) = \alpha \leq 0$, for the optimal state, we have $y(t) \leq 0$ for all $t \geq 0$. (Since, otherwise, instead of increasing the state above zero, it is better to switch off the control.) Moreover, for the optimal control, we have $u(t) \geq 0$. (Since, otherwise, instead of decreasing the state, it is also better to switch off the control). Hence, it suffices to consider the feasible controls $u(t) \geq 0$ that satisfy the moment inequality

$$\int_0^T \frac{g(\tau)}{F(\tau)}\,u(\tau)\,d\tau \leq -\alpha. \tag{2.9}$$

Due to the choice of λ, for the state $\hat{y}$ generated by $\hat{u}$, we have $\hat{y}(T) = 0$. For $t \in [0,\ T]$, consider

$$B(t) = \int_0^t \frac{g(\tau)}{F(\tau)}\,\hat{u}(\tau)\,d\tau. \tag{2.10}$$

Then, $B(0) = 0$ and B is increasing. Hence, also the function $[\alpha + B(t)]$ is increasing. We have $B(0) + \alpha < 0$ and $B(T) + \alpha = 0$. Thus, there exists a unique point

$$t_0 = \min\{t \in [0,\ T] : \alpha + B(t) = 0\},$$

and we have $t_0 \in (0,\ T]$. We have $B(t_0) = B(T)$ and B is increasing. This implies that for all $t \in [t_0,\ T]$, we have $B(t) = -\alpha$. On account of the definition of B as an integral, this is only possible if for all $t \in [t_0,\ T]$, we have $\hat{u}(t) = 0$. This implies that for all $t \in [t_0,\ T]$, we have

$$-1 - \gamma\,\frac{g(t)\,H(t)}{F(t)} + \lambda \frac{g(t)}{F(t)} \leq 0. \tag{2.11}$$

By (2.6), we have

$$t_0 = \min\{t \in [0,\ T] : \hat{y}(t) = 0\}.$$

Since $\hat{y}(t) = F(t)\,[\alpha + B(t)]$, for $t < t_0$, we have $\hat{y}(t) < 0$. Since for $t \geq t_0$, we have $\hat{u}(t) = 0$, this implies that $\hat{y}(t) = 0$ for all $t \geq t_0$.

Since $\hat{u} \geq 0$ and $\hat{y} \leq 0$, for the objective function, we have

$$J(\hat{u}) = \int_0^T \frac{1}{2}\,(\hat{u}(t))^2 + \hat{u}(t) - \gamma\, F(t) \left[\alpha + \int_0^t \hat{u}(\tau)\, \frac{g(\tau)}{F(\tau)}\, d\tau\right] dt.$$

Integration by parts yields (since $B(T) = -\alpha$)

$$\begin{aligned}
J(\hat{u}) &= \int_0^T \frac{1}{2}\,(\hat{u}(t))^2 + \hat{u}(t)\, dt - \gamma\, H(s) \left[\alpha + \int_0^s \hat{u}(\tau)\, \frac{g(\tau)}{F(\tau)}\, d\tau\right] |_{s=0}^{T} \\
&+ \gamma \int_0^T H(t)\, \hat{u}(t)\, \frac{g(t)}{F(t)}\, dt \\
&= \int_0^T \frac{1}{2}\,(\hat{u}(t))^2 + \hat{u}(t)\, dt - \gamma\, H(t_0)\, (\alpha + B(T)) \\
&+ \gamma \int_0^T \hat{u}(\tau)\, \frac{H(\tau)\, g(\tau)}{F(\tau)}\, d\tau \\
&= \int_0^T \frac{1}{2}\,(\hat{u}(t))^2 + \hat{u}(t) \left[1 + \gamma\, \frac{H(t)\, g(t)}{F(t)}\right] dt.
\end{aligned}$$

Let $\delta \in L^2(0, T)$ be given. We use δ as a perturbation of the control. To make sure that the terminal condition remains valid, we assume that

$$\int_0^T \delta(\tau)\, \frac{g(\tau)}{F(\tau)}\, d\tau = 0. \tag{2.12}$$

Since the optimal control must increase the values of the corresponding trajectory to zero, it can only have positive values. Therefore, we assume that for $t \in [0, T]$, we have $\hat{u}(t) + \delta(t) \geq 0$. Thus, for $t \in [0, t_0]$, we have $\operatorname{sign}(\hat{u}(t) + \delta(t)) = 1$, and for $t \geq t_0$, we have $\delta(t) \geq 0$. Then, we have

$$\begin{aligned}
J(\hat{u} + \delta) &= \int_0^T \frac{1}{2}\,(\hat{u}(t) + \delta(t))^2 + (\hat{u}(t) + \delta(t)) \left[\operatorname{sign}(\hat{u}(t) + \delta(t)) + \gamma\, \frac{H(t)\, g(t)}{F(t)}\right] dt \\
&= J(\hat{u}) + \int_0^T \frac{1}{2}\delta(t)^2\, dt + \int_0^{t_0} \delta(t) \left[\hat{u} + 1 + \gamma\, \frac{H(t)\, g(t)}{F(t)}\right] dt \\
&+ \int_{t_0}^T \delta(t) \left[\operatorname{sign}(\delta(t)) + \gamma\, \frac{H(t)\, g(t)}{F(t)}\right] dt \\
&= J(\hat{u}) + \int_0^T \frac{1}{2}\delta(t)^2\, dt + \int_0^{t_0} \delta(t)\, \lambda\, \frac{g(t)}{F(t)}\, dt
\end{aligned}$$

$$
\begin{aligned}
&+ \int_{t_0}^{T} \delta(t) \left[\operatorname{sign}(\delta(t)) + \gamma \, \frac{H(t)\, g(t)}{F(t)} \right] dt \\
&= J(\hat{u}) + \int_0^T \frac{1}{2}\delta(t)^2 \, dt + \int_0^T \delta(t)\, \lambda \, \frac{g(t)}{F(t)} \, dt \\
&+ \int_{t_0}^{T} \delta(t) \left[-\lambda \, \frac{g(t)}{F(t)} + \operatorname{sign}(\delta(t)) + \gamma \, \frac{H(t)\, g(t)}{F(t)} \right] dt \\
&= J(\hat{u}) + \int_0^T \frac{1}{2}\delta(t)^2 \, dt \\
&+ \int_{t_0}^{T} \delta(t) \left[1 + \gamma \, \frac{H(t)\, g(t)}{F(t)} - \lambda \, \frac{g(t)}{F(t)} \right] dt \\
&\geq J(\hat{u}),
\end{aligned}
$$

where the last step follows with (2.11). Thus, $\hat{u}$ is the minimizer of J among all controls that generate states with $y(T) = 0$. This shows the assertion. □

The question remains: Do we have $t_0 < T$ if γ is sufficiently large?
Let $t_1 \in (0, T)$ be given such that

$$
-\alpha - \frac{F(t_1)}{g(t_1)} \int_0^{t_1} \frac{g^2}{F^2} \, dt + \int_0^{t_1} \frac{g}{F} \, dt > 0. \tag{2.13}
$$

Note that H is strictly increasing, and hence we have the inequality

$$
\int_0^{t_1} (H(t_1) - H(t)) \, \frac{g^2}{F^2} \, dt > 0.
$$

Define the number

$$
\gamma(t_1) = \frac{-\alpha - \frac{F(t_1)}{g(t_1)} \int_0^{t_1} \frac{g^2}{F^2} \, dt + \int_0^{t_1} \frac{g}{F} \, dt}{\int_0^{t_1} (H(t_1) - H(t)) \frac{g^2}{F^2} \, dt}. \tag{2.14}
$$

Then, we have $\gamma(t_1) > 0$. Define the number

$$
\lambda_1 = \frac{-\alpha + \int_0^{t_1} \frac{g}{F} \, dt + \gamma(t_1) \int_0^{t_1} H \frac{g^2}{F^2} \, dt}{\int_0^{t_1} \frac{g^2}{F^2} \, dt}. \tag{2.15}
$$

The definition of λ_1 implies the equation

$$
\lambda_1 \int_0^{t_1} \frac{g^2}{F^2} \, dt = -\alpha + \int_0^{t_1} \frac{g}{F} \, dt + \gamma(t_1) \int_0^{t_1} H \frac{g^2}{F^2} \, dt. \tag{2.16}
$$

Moreover, due to the definition of $\gamma(t_1)$, we have

$$1 + \frac{g(t_1)\,H(t_1)}{F(t_1)}\,\gamma(t_1) = \frac{\int_0^{t_1} H\frac{g^2}{F^2}\,dt + \frac{g(t_1)\,H(t_1)}{F(t_1)}\left[\alpha - \int_0^{t_1}\frac{g}{F}\,dt\right]}{\int_0^{t_1} H\frac{g^2}{F^2}\,dt - H(t_1)\int_0^{t_1}\frac{g^2}{F^2}\,dt}.$$

In addition, the definition of λ_1 and $\gamma(t_1)$ implies

$$\lambda_1\,\frac{g(t_1)}{F(t_1)} = \frac{-\frac{g(t_1)}{F(t_1)}\left[\alpha - \int_0^{t_1}\frac{g}{F}\,dt\right] + \frac{g(t_1)}{F(t_1)}\gamma(t_1)\int_0^{t_1} H\frac{g^2}{F^2}\,dt}{\int_0^{t_1}\frac{g^2}{F^2}\,dt}$$

$$= \frac{-\frac{g(t_1)}{F(t_1)}\left[\alpha - \int_0^{t_1}\frac{g}{F}\,dt\right] + \left[\frac{\int_0^{t_1}\frac{g^2}{F^2}\,dt + \frac{g(t_1)}{F(t_1)}\left[\alpha - \int_0^{t_1}\frac{g}{F}\,dt\right]}{\int_0^{t_1} H\frac{g^2}{F^2}\,dt - H(t_1)\int_0^{t_1}\frac{g^2}{F^2}\,dt}\right]\int_0^{t_1} H\frac{g^2}{F^2}\,dt}{\int_0^{t_1}\frac{g^2}{F^2}\,dt}$$

$$= \frac{1}{\int_0^{t_1}\frac{g^2}{F^2}\,dt}\left[\frac{\int_0^{t_1} H\frac{g^2}{F^2}\,dt}{\int_0^{t_1} H\frac{g^2}{F^2}\,dt - H(t_1)\int_0^{t_1}\frac{g^2}{F^2}\,dt} - 1\right]\frac{g(t_1)}{F(t_1)}\left[\alpha - \int_0^{t_1}\frac{g}{F}\,dt\right]$$

$$+ \frac{\int_0^{t_1} H\frac{g^2}{F^2}\,dt}{\int_0^{t_1} H\frac{g^2}{F^2}\,dt - H(t_1)\int_0^{t_1}\frac{g^2}{F^2}\,dt}$$

$$= \frac{\frac{g(t_1)\,H(t_1)}{F(t_1)}\left[\alpha - \int_0^{t_1}\frac{g}{F}\,dt\right] + \int_0^{t_1} H\frac{g^2}{F^2}\,dt}{\int_0^{t_1} H\frac{g^2}{F^2}\,dt - H(t_1)\int_0^{t_1}\frac{g^2}{F^2}\,dt}.$$

Hence, we have

$$\lambda_1\,\frac{g(t_1)}{F(t_1)} = 1 + \gamma(t_1)\,\frac{g(t_1)\,H(t_1)}{F(t_1)}.$$

Assume that g is continuously differentiable and that we have

$$g'(t) \le f(t)\,g(t). \tag{2.17}$$

Assumption (2.17) implies that the function $\frac{g}{F}$ is decreasing.

Since the function $(\lambda_1 - \gamma(t_1)\,H)$ is decreasing and $\frac{g}{F} > 0$, this implies that also the product

$$\frac{g}{F}\,(\lambda_1 - \gamma(t_1)\,H)$$

is decreasing as a function of time.

Then, the optimal control $\hat{u}$ as defined in (2.7) (with $\lambda = \lambda_1$ and $\gamma = \gamma(t_1)$) is decreasing, $\hat{u}(t_1) = 0$, and the support of the optimal control $\hat{u}$ is contained in $[0, t_1]$. With λ_1 defined as in (2.15), Eq. (2.16) holds. This implies that the optimal control $\hat{u}$ as defined in (2.7) satisfies (2.8). Thus, we have shown the following statement:

If (2.17) holds, for all $t_0 \in (0, T)$ such that (2.13) holds (with $t_1 = t_0$), there is a weight $\gamma > 0$ such that the support of the corresponding optimal control is contained in $[0, t_0]$.

Note that in Example 1, we have $f(t) = 1$ and $g(t) = \exp(t) = g'(t)$, and hence (2.17) holds. This explains why in the first example, for sufficiently large values of T, no terminal constraint is necessary. As a second example, for the constant function $g(t) = 1$, (2.17) also holds.

3 General Results in Hilbert Spaces

In this section, we study optimal control problems in a Hilbert space setting. In this way, we obtain results that we can apply to systems that are governed by partial differential equations. Let X and U be Hilbert spaces with the inner products $\langle \cdot, \cdot \rangle_X$ and $\langle \cdot, \cdot \rangle_U$ and the corresponding norms $\| \cdot \|_X$ and $\| \cdot \|_U$, respectively. We use $T > 0$ to denote the terminal time of our optimal control problems. The space X contains the current state, and the space U is used as a framework for the control functions in $L^2(0, T; U)$.

Let $A : \mathcal{D}(A) \subset X \to X$ be the generator of a strongly continuous semigroup, and let B denote an admissible control operator. As in [16, Proposition 4.2.5.], we consider control systems of the form

$$\begin{cases} x' + Ax = Bu, \\ x(0) = x_0, \end{cases} \tag{3.1}$$

where $x_0 \in X$ is a given initial state. For all $u \in L^2(0, T; U)$, the Cauchy problem (3.1) has a unique solution $x \in C([0, T]; X)$ (see [12]).

3.1 Exact Controllability

Assume that (3.1) is *exactly controllable using L^2-controls* in time $t_0 > 0$, that is, there exists a constant $C_1 > 0$ such that for all initial states $x_0 \in X$ and all terminal states $x_1 \in X$, there is a control $u \in L^2(0, t_0; U)$ such that the solution $x \in C([0, t_0]; X)$ of (3.1) satisfies

$$\begin{cases} x(t_0) = x_1. \\ \|u\|_{L^2(0,t_0;U)} \le C_1(\|x_0\|_X + \|x_1\|_X). \end{cases} \tag{3.2}$$

Let a desired state $x_d \in X$ be given. Due to the exact controllability assumption, there exists a control $u_{\text{exact}} \in L^2(0, t_0; U)$ such that the solution $x_{\text{exact}} \in C([0, t_0]; X)$ of (3.1) satisfies

$$x_{\text{exact}}(t_0) = x_d. \tag{3.3}$$

We assume that x_d is a *holdable state* in the sense that we can extend u_{exact} to the time interval $[0, T]$ by a constant control u_d on $[t_0, T]$ such that for the corresponding state, for all $t \in (t_0, T)$, we have the equation $x_{\text{exact}}(t) = x_d$ and $u_{\text{exact}}(t) = u_d$. Thus, on the time interval (t_0, T), we have $A\, x_{\text{exact}} = A\, x_d = B\, u_{\text{exact}}$.

3.2 An Optimal Control Problem with Max-Norm Penalization

First, we consider a tracking term with the maximum-norm. For systems that are exactly controllable, the optimal control steers the system to the desired state after the prescribed time t_0.

For $\gamma > 0$, we consider the following optimization problem:

$$\mathbf{P}(T, \gamma) \begin{cases} \min\limits_{u \in L^2(0,t;\, U)} \; \frac{1}{2}\, \|u - u_d\|^2_{L^2(0,T;U)} + \gamma\, \max_{t \in [t_0,\, T]} \|x(s) - x_d\|_X \\ \text{subject to} \\ x' + Ax = Bu, \;\; x(0) = x_0. \end{cases}$$

In problem $\mathbf{P}(T, \gamma)$, the end condition $x(t) = x_d$ does not appear. Note that problem $\mathbf{P}(T, \gamma)$ has a unique solution.

Our goal is to show that, due to the property of exact controllability using L^2-controls of the system, for γ sufficiently large, the optimal state x_T satisfies the condition

$$x_T(t) = x_d$$

for all $t \in [t_0, T]$. A precise statement is given in the following theorem:

Theorem 2 *Assume that $T > t_0$, and that the system (3.1) is exactly controllable. If $\gamma > 0$ is sufficiently large, for all $t \in [t_0, T]$, the solution (u_T, x_T) of problem $\mathbf{P}(T, \gamma)$ satisfies the equation*

$$x_T(t) = x_d.$$

Proof An application of the Direct Method of the Calculus of Variations shows that a solution of $\mathbf{P}(T, \gamma)$ exists. The strict convexity of the control cost $\frac{1}{2}\, \|\cdot\|^2_{L^2(0,T;U)}$

implies that the solution of $\mathbf{P}(T, \gamma)$ is uniquely determined. Choose

$$\gamma > C_1 \, \|u_{\text{exact}} - u_d\|_{L^2(0,t_0;U)}. \tag{3.4}$$

Similarly, as in [8], consider the optimal control problem

$$\mathbf{Q}(T,\gamma) \begin{cases} \min\limits_{u \in L^2(0,T;\, U)} \frac{1}{2} \, \|u - u_d\|^2_{L^2(0,T;U)} + \gamma \, \|x(t_0) - x_d\|_X \\ \text{subject to} \\ x' + Ax = Bu, \; x(0) = x_0. \end{cases}$$

Let $(u^*, \, x^*)$ denote the solution of $\mathbf{Q}(T,\gamma)$. Now, similarly as in Theorem 1 in [8], we show that $x^*(t_0) = x_d$ by an indirect proof.

Suppose that $x^*(t_0) \neq x_d$. Then, the objective functional of $\mathbf{Q}(T,\gamma)$ is differentiable at $(u^*, \, x^*)$, and the necessary optimality conditions imply

$$\int_0^T \langle u^* - u_d, v\rangle_U \, dt + \gamma \, \frac{\langle x^*(t_0) - x_d, y\rangle_X}{\|x^*(t_0) - x_d\|_X} = 0 \tag{3.5}$$

for all $v \in L^2(0, t_1; U)$, where y solves

$$y' + Ay = Bv, \; y(0) = 0.$$

Due to the exact controllability of the system, we can choose a control $\tilde{v} \in L^2(0, t_0; U)$ such that for the corresponding state $\tilde{y}$, we have

$$\tilde{y}(t_0) = \frac{x^*(t_0) - x_d}{\|x^*(t_0) - x_d\|_X}$$

and

$$\|\tilde{v}\|_{L^2(0,t_0;U)} \leq C_1. \tag{3.6}$$

We extend $\tilde{v}$ to an element of $L^2(0, T; U)$ by the definition $\tilde{v}(s) = 0$ for all $s \in (t_0, \, T)$. Then, the necessary optimality condition yields the equation

$$\int_0^T \langle u^* - u_d, \tilde{v}\rangle_U \, dt + \gamma \, \frac{\langle x^*(t_0) - x_d, \, \frac{x^*(t_0) - x_d}{\|x^*(t_0) - x_d\|_X}\rangle_X}{\|x^*(t_0) - x_d\|_X} = 0. \tag{3.7}$$

This implies the equation

$$\left| \int_0^T \langle u^* - u_d, \tilde{v}\rangle_U \, dt \right| = \gamma. \tag{3.8}$$

On the other hand, we have the inequality

$$\left| \int_0^T \langle u^* - u_d, \tilde{v} \rangle_U \, dt \right| \leq \|u^* - u_d\|_{L^2(0,T;U)} \, \|\tilde{v}\|_{L^2(0,T;U)}.$$

Since the control u_{exact} is feasible for $\mathbf{Q}(T, \gamma)$, we have the inequality

$$\frac{1}{2} \|u^* - u_d\|^2_{L^2(0,T;U)} \leq \frac{1}{2} \|u_{\text{exact}} - u_d\|^2_{L^2(0,T;U)} + \gamma \, \|y_{\text{exact}}(t_0) - y_d\|_X$$

$$= \frac{1}{2} \|u_{\text{exact}} - u_d\|^2_{L^2(0,T;U)}.$$

Hence,

$$\|u^* - u_d\|_{L^2(0,T;U)} \leq \|u_{\text{exact}} - u_d\|_{L^2(0,T;U)}.$$

Moreover, (3.6) implies

$$\|\tilde{v}\|_{L^2(0,T;U)} \leq C_1.$$

Hence, (3.8) implies

$$\gamma \leq C_1 \, \|u_{\text{exact}} - u_d\|_{L^2(0,T;U)},$$

which is a contradiction to (3.4). Thus, we have shown that $x^*(t_0) = x_d$. This implies that for $s \in (t_0, T]$, we have $u^*(s) = u_d$ and $x^*(s) = x_d$.

Let v_Q denote the optimal value of $\mathbf{Q}(T, \gamma)$ and v_P denote the optimal value of $\mathbf{P}(T, \gamma)$. Then, the definition of the corresponding objective functionals implies the inequality

$$v_Q \leq v_P.$$

Since the control u^* is feasible for $\mathbf{P}(T, \gamma)$, we also have the inequality

$$v_P = \frac{1}{2} \|u_T - u_d\|^2_{L^2(0,T;U)} + \gamma \max_{t \in [t_0, T]} \|x_T(s) - x_d\|_X$$

$$\leq \frac{1}{2} \|u^* - u_d\|^2_{L^2(0,T;U)} + \gamma \max_{t \in [t_0, T]} \|x^*(s) - x_d\|_X$$

$$= \frac{1}{2} \|u^* - u_d\|^2_{L^2(0,t_0;U)} = v_Q.$$

Thus, we have $v_P = v_Q$, and (u^*, x^*) is an optimal control/state pair for $\mathbf{Q}(T, \gamma)$. Since the solution is unique, this implies the assertion. □

3.3 An Optimal Control Problem for Nodal Profile Exactly Controllable Systems

Motivated by application problems in the operation of gas pipelines, the exact controllability of nodal profiles has been introduced in [9], see also [11]. The assumption of exact controllability of nodal profiles also allows to derive a result about the exactness of an L^2-norm penalty term.

Let a Hilbert space Z, $t_0 \in (0, T)$ and a linear map $\Pi : L^2(0, T; X) \to L^2(t_0, T; Z)$ be given. In the applications, typically, Π will be some trace operator, for example, the boundary trace of the system state restricted to the time interval $[t_0, T]$, see [9].

Assume that (3.1) is *nodal profile exactly controllable using L^2-controls* in time $t_0 > 0$, that is, there exists a constant $C_1 > 0$ such that for all initial states $x_0 \in X$ and all nodal profiles $z \in L^2(t_0, T; Z)$, there is a control $u \in L^2(0, T; U)$ such that the solution $x \in C([0, T]; X)$ of (3.1) satisfies for all $t \in [t_0, T]$

$$\begin{cases} \Pi x(t) = z(t), \\ \|u\|_{L^2(0,T;U)} \le C_1 \left(\|x_0\|_X + \|z\|_{L^2(t_0,T;Z)}\right). \end{cases} \tag{3.9}$$

Remark 1 The exact boundary controllability of nodal profile for hyperbolic systems is discussed in [11].

For $\gamma > 0$, we consider the following optimization problem:

$$\mathbf{S}(T, \gamma) \begin{cases} \min\limits_{u \in L^2(0,t;\, U)} \frac{1}{2} \|u - u_d\|^2_{L^2(0,T;U)} + \gamma \sqrt{\int_{t_0}^{T} \|\Pi x(s) - \Pi x_d\|_Z^2 \, ds} \\ \text{subject to} \\ x' + Ax = Bu, \; x(0) = x_0, \end{cases}$$

where as before, $x_d \in X$ is the desired holdable state. In problem $\mathbf{S}(T, \gamma)$, the end condition $x(t) = x_d$ does not appear. Note that problem $\mathbf{S}(T, \gamma)$ has a unique solution.

Remark 2 Optimization problems of a similar structure with a differentiable tracking term have been considered in [7] and [6].

Due to the nodal profile exact controllability assumption, there exists a control $v_{\text{exact}} \in L^2(0, t_0; U)$ such that the solution $p_{\text{exact}} \in C([0, t_0]; X)$ of (3.1) satisfies

$$\Pi p_{\text{exact}}(t) = \Pi x_d \tag{3.10}$$

for all $t \in [t_0, T]$.

Our goal is to show that, due to the property of nodal profile exact controllability using L^2-controls of the system, for γ sufficiently large, the optimal state x_T satisfies the condition

$$\Pi x_T(t) = \Pi x_d$$

for all $t \in [t_0, T]$. In the application in supply systems, this means that on the time interval $[t_0, T]$, the nodal profile that is desired by the customer is attained exactly. A precise statement is given in the following theorem:

Theorem 3 *Assume that $T > t_0$, and that the system (3.1) is nodal profile exactly controllable. If $\gamma > C_1 \|u_{\text{exact}} - u_d\|_{L^2(0,t_0;U)}$, for all $t \in [t_0, T]$, the solution (u_T, x_T) of problem $\mathbf{S}(T, \gamma)$ satisfies the equation*

$$\Pi x_T(s) = \Pi x_d$$

for all $s \in [t_0, T]$.

Proof An application of the Direct Method of the Calculus of Variations shows that a solution of $\mathbf{S}(T, \gamma)$ exists. The strict convexity of the control cost $\frac{1}{2}\|\cdot\|^2_{L^2(0,T;U)}$ implies that the solution of $\mathbf{S}(T, \gamma)$ is uniquely determined. Choose

$$\gamma > C_1 \|v_{\text{exact}} - u_d\|_{L^2(0,t_0;U)}. \tag{3.11}$$

Suppose that there exists $\tau \in [t_0, T]$ such that $\Pi x^*(\tau) \neq \Pi x_d$. Then, $\|\Pi x^* - \Pi x_d\|_{L^2(t_0,T;Z)} \neq 0$. Hence, the objective functional of $\mathbf{S}(T, \gamma)$ is differentiable in (u^*, x^*), and the necessary optimality conditions imply

$$\int_0^T \langle u^* - u_d, v\rangle_U \, dt + \gamma \int_{t_0}^T \frac{\langle \Pi x^*(\tau) - \Pi x_d, \Pi y(\tau)\rangle_Z}{\|\Pi x^* - \Pi x_d\|_{L^2(t_0,T;Z)}} \, d\tau = 0 \tag{3.12}$$

for all $v \in L^2(0, T; U)$, where y solves

$$y' + Ay = Bv, \ \ y(0) = 0.$$

Due to the nodal profile exact controllability of the system, we can choose a control $\tilde{v} \in L^2(0, t_0; U)$ such that for the corresponding state $\tilde{y}$, we have for all $\tau \in [t_0, T]$

$$\Pi \tilde{y}(\tau) = \frac{\Pi x^*(\tau) - \Pi x_d}{\|\Pi x^* - \Pi x_d\|_{L^2(t_0,T;Z)}}$$

and

$$\|\tilde{v}\|_{L^2(0,T;U)} \leq C_1.$$

Then, the necessary optimality condition (3.12) yields the equation

$$\int_0^T \langle u^* - u_d, \tilde{v}\rangle_U \, dt + \gamma \int_{t_0}^T \frac{\langle \Pi x^*(\tau) - \Pi x_d, \frac{\Pi x^*(\tau) - \Pi x_d}{\|\Pi x^* - \Pi x_d\|_{L^2(t_0,T;Z)}}\rangle_Z}{\|\Pi x^* - \Pi x_d\|_{L^2(t_0,T;Z)}} \, d\tau = 0. \tag{3.13}$$

This implies the equation

$$\left| \int_0^T \langle u^* - u_d, \tilde{v}\rangle_X \, dt \right| = \gamma. \tag{3.14}$$

On the other hand, we have the inequality

$$\left| \int_0^T \langle u^* - u_d, \tilde{v}\rangle_U \, dt \right| \leq \|u^* - u_d\|_{L^2(0,T;U)} \, \|\tilde{v}\|_{L^2(0,T;U)}.$$

Since the control v_{exact} is feasible for $\mathbf{S}(T, \gamma)$, we have the inequality

$$\begin{aligned} \frac{1}{2}\|u^* - u_d\|^2_{L^2(0,T;U)} &\leq \frac{1}{2}\|v_{\text{exact}} - u_d\|^2_{L^2(0,T;U)} + \gamma \int_{t_0}^T \|\Pi p_{\text{exact}}(\tau) - \Pi x_d\|_Z \, d\tau \\ &= \frac{1}{2}\|v_{\text{exact}} - u_d\|^2_{L^2(0,T;U)}. \end{aligned}$$

Hence,

$$\|u^* - u_d\|_{L^2(0,T;U)} \leq \|v_{\text{exact}} - u_d\|_{L^2(0,T;U)}.$$

Moreover, we have

$$\|\tilde{v}\|_{L^2(0,T;U)} \leq C_1.$$

Hence, (3.14) implies

$$\gamma \leq C_1 \, \|v_{\text{exact}} - u_d\|_{L^2(0,T;U)},$$

which is a contradiction to (3.11). Thus, we have shown that $\Pi x^* = \Pi x_d$ on $[t_0, T]$. This implies the assertion. □

3.4 An Optimal Control Problem with L^1-Norm Tracking Term

In this section, we present a result about the finite-time turnpike structure of the optimal state and the optimal control that we have shown under the assumption of exact controllability (3.2) for an optimal control problem with an L^1-norm tracking term with a singular weight in the objective function.

For $\gamma > 0$, we consider the following optimal control problem $\mathbf{R}(T, \gamma)$ with L^1-norm tracking term:

$$\mathbf{R}(T,\gamma)\begin{cases} \min\limits_{u\in L^2(0,T;\,U)} \ \frac{1}{2}\,\|u-u_d\|^2_{L^2(0,T;U)} + \gamma \int_{t_0}^{T} \frac{1}{s-t_0}\|x(s)-x_d\|_X\, ds \\ \text{subject to} \\ x' + Ax = Bu, \ x(0) = x_0. \end{cases}$$

In problem $\mathbf{R}(T, \gamma)$, the end condition $x(T) = x_d$ does not appear. Problem $\mathbf{R}(T, \gamma)$ has a unique solution.

Our goal is to show that, due to the property of exact controllability using L^2-controls of the system, for γ and T sufficiently large, the optimal state x_T for $\mathbf{R}(T, \gamma)$ satisfies the condition

$$x_T(t) = x_d$$

for all $t \in (t_0, T]$. A precise statement is given in the following theorem:

Theorem 4 *Assume that $T > t_0$, and that the system (3.1) is exactly controllable. If $\gamma > 0$, the solution (u_T, x_T) of problem $\mathbf{R}(T, \gamma)$ satisfies the equation*

$$x_T(t) = x_d$$

for all $t \in [t_0, T]$.

Proof Since u_{exact} is a feasible control for $\mathbf{R}(T, \gamma)$, evaluating the objective function of $\mathbf{R}(T, \gamma)$ at u_{exact} yields the inequality

$$\|u_T - u_d\|^2_{L^2(0,T;U)} \leq \|u_T - u_d\|^2_{L^2(0,t_0;U)} + 2\gamma \int_{t_0}^{T} \frac{1}{s-t_0}\|x_{\text{exact}}(s) - x_d\|_X\, ds \tag{3.15}$$

$$= \|u_{\text{exact}} - u_d\|^2_{L^2(0,t_0;U)}.$$

An application of the Direct Method of the Calculus of Variations shows that a solution of $\mathbf{R}(T, \gamma)$ exists. For the optimal control/state pair, we use the notation (u_T, x_T).

If there exists $\hat{t} \in (0, T)$ with $x_T(\hat{t}) = x_d$, the optimal way to continue the control for $s \in (\hat{t}, T]$ is with (u_d, x_d), and hence for all $s \in (\hat{t}, T]$, we have $x_T(s) = x_d$.

Suppose that there exists a number $t_1 \in (t_0, T]$ such that $x_T(t_1) \neq x_d$. Then, for all $t \in [t_0, t_1)$, we also have $x_T(t) \neq x_d$. In particular, for all $t \in [t_0, t_1]$, we have $\|x_T(t) - x_d\|_X > 0$. Since x_T is continuous, this implies that

$$\inf_{t \in [t_0, t_1]} \|x_T(t) - x_d\|_X = \varepsilon > 0.$$

This implies

$$\int_{t_0}^{t_1} \frac{1}{s - t_0} \|x(s) - x_d\| \, ds \geq \varepsilon \int_{t_0}^{t_1} \frac{1}{s - t_0} = \infty.$$

Hence, x_T cannot be optimal, and this is a contradiction.

□

4 Examples

In this section, we present some examples to illustrate our results about the finite-time turnpike phenomenon. We start with one example with a system that is governed by an ordinary differential equation, and then we present examples with partial differential equations.

Example 2 Let us first return to Example 1. Here, we present numerical results that illustrate the numerical solution for the discretized optimal control problem, where for $T = 2$, the interval $[0, 2]$ has been replaced with a grid of 201 equidistant points and the ordinary differential equation has been replaced by a discrete-time system with the Euler backward discretization.

The resulting optimization problem has been solved numerically with a standard method from Matlab. To improve the performance, in the numerical experiments, the constraints $u \geq 0$ and $y \leq 0$ have been included in the problem. (As shown in Example 1, they do not change the solution.) The numerical results are presented in Fig. 1 for $\gamma = \frac{1}{2}$, Fig. 2 for $\gamma = 1$, and Fig. 3 for $\gamma = 2$.

Now, we present examples of optimal control problems where Theorem 2 or Theorem 4 is applicable. These theorems assume that the system is exactly controllable.

Example 3 Now, we consider a problem of optimal torque control for an Euler–Bernoulli beam. Let $y_0 \in H^2(0, 1)$ and $y_1 \in H^1(0, 1)$ be given. We study the

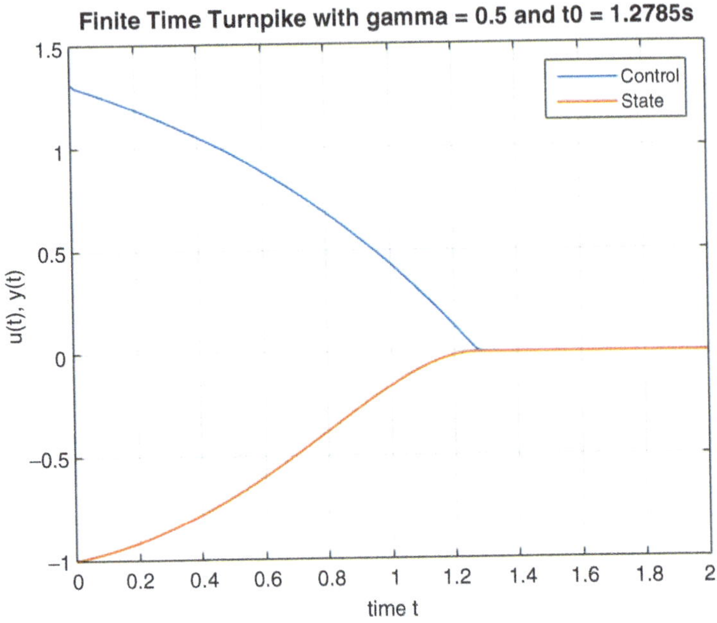

Fig. 1 The figure shows the optimal control and the optimal state as approximate solutions of problem $(\mathbf{OC})_T$ for $T = 2$ and $\gamma = \frac{1}{2}$ defined in Example 2

following optimal control problem:

$$\begin{cases} \min\limits_{u\in L^2(0,T)} \quad \frac{1}{2}\|u^2(t)\|^2\,dt + \gamma\ \max_{t\in[t_0,\,T]}\|y(t,\,\cdot)\|_{L^2(0,1)} \text{ subject to} \\ y(0,x) = y_0(x),\ \ y_t(0,x) = y_1(x),\ \ x \in (0,1) \\ y(t,0) = 0,\ \ y_{xx}(t,0) = u(t),\ t \in (0,T) \\ y(t,\,1) = y_{xx}(t,\,1) = 0, \\ y_{tt}(t,x) = -y_{xxxx}(t,x),\ \ (t,x) \in (0,T) \times (0,1). \end{cases}$$

We have $U = L^2(0,1)$ and $X = L^2(0,1)$. Note that the Euler–Bernoulli beam is exactly controllable in arbitrarily short times (see [16, Example 11.2.8]), so in this case, $t_0 > 0$ can be chosen arbitrarily small. Theorem 2 implies that if γ is chosen sufficiently large, the beam is steered to a position of rest in the time $t_0 > 0$.

Example 4 Consider the problem of optimal Neumann boundary control of the wave equation. Define $Q = (0,T) \times (0,1)$. Here, we have $U = L^2(0,1)$, $X = L^1(0,1)$.

Let $y_d \in X$ and $u_d \in U$ be given. Consider the optimal control problem

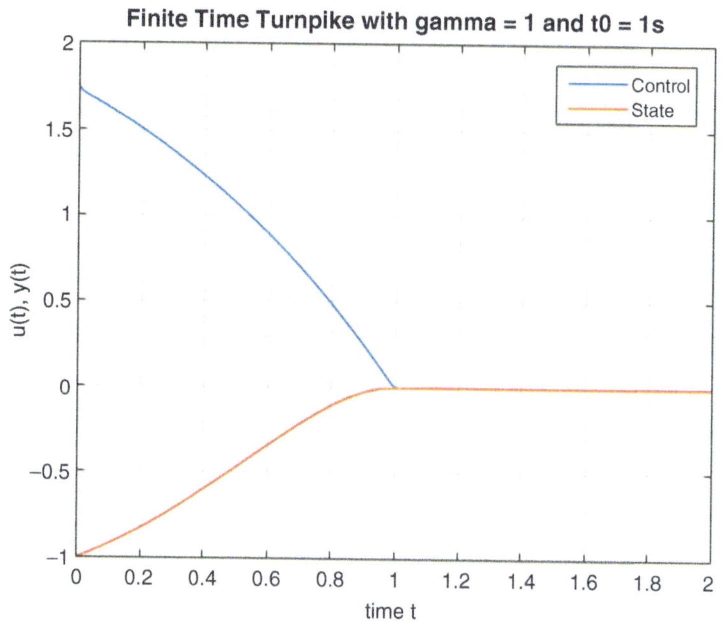

Fig. 2 The figure shows the optimal control and the optimal state as approximate solutions of problem $(\mathbf{OC})_T$ for $T = 2$ and $\gamma = 1$ defined in Example 2

$$\begin{cases} \min\limits_{u \in U} \frac{1}{2} \int\limits_0^T (u(t) - u_d)^2 \, dt + \int\limits_2^T \frac{1}{t-2} \int\limits_0^1 |y(t,x) - y_d| \, dx \, dt \text{ subject to} \\ y(0,x) = 0, \ \ y_t(0,x) = 0, \ \ x \in (0,1) \\ y(t,0) = 0, \ \ y_x(t,1) = u(t), \ t \in (0,T) \\ y_{tt}(t,x) - y_{xx}(t,x) = 0, \ \ (t,x) \in Q. \end{cases}$$

Our results show that the solution has a turnpike structure as described in Theorem 4. The optimal control problem is similar to the Neumann optimal boundary control problem with a differentiable objective function considered in [10].

Now, we present an example where Theorem 3 is applicable, which assumes that the system is nodal profile exactly controllable.

Example 5 Now, we consider a problem or optimal control where Theorem 3 is applicable. The problem is similar as in [7], but in the tracking term, instead of the squared L^2-norm we take the L^2-norm. The motivation for this type of problem where the boundary trace of the state is driven to a desired profile comes from the operation of networks of gas pipelines, where the aim is to satisfy customer demands in an optimal way.

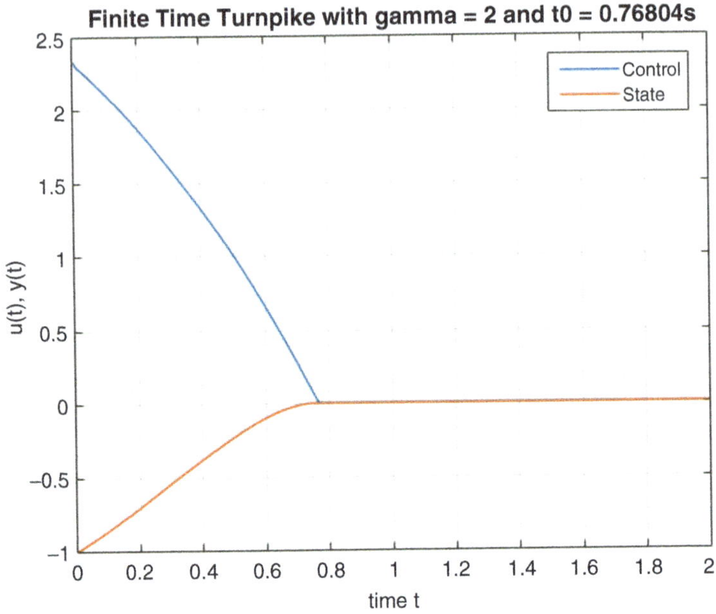

Fig. 3 The figure shows the optimal control and the optimal state as approximate solutions of problem $(\mathbf{OC})_T$ for $T = 2$ and $\gamma = 2$ defined in Example 2

We consider a 2×2 system in diagonal form. Let a length $L > 0$ and a time interval $[0, T]$ be given. Let d_- and d_+ be real numbers such that

$$d_- < 0 < d_+.$$

Define the diagonal matrix

$$D = \begin{pmatrix} d_+ & 0 \\ 0 & d_- \end{pmatrix}.$$

For all $x \in [0, L]$, let $M(x)$ denote a 2×2 matrix that depends continuously on x. Assume that for all $x \in [0, L]$, the matrix $M(x)$ is positive semi-definite. Let $\eta_0 \leq 0$ be a real number.

Consider the linear hyperbolic partial differential equation

$$r_t + D\, r_x = \eta_0\, M\, r, \tag{4.1}$$

where for $x \in (0, L)$ and $t \in (0, T)$, the state is given by $r(t, x) = \begin{pmatrix} r_+(t, x) \\ r_-(t, x) \end{pmatrix}$.

Let real numbers R_+^d and R_-^d be given. To obtain an initial boundary value problem, in addition to (4.1), we consider the initial condition $r(0, x) = 0$ for

$x \in (0, L)$ at the time $t = 0$, and for $t \in (0, T)$ the Dirichlet boundary conditions $r_+(t, 0) = u_+(t)$, $r_-(t, L) = R_-^d$, with a boundary control u_+ in $L^2(0, T)$. The resulting initial boundary value problem

$$\begin{cases} r(0, x) = 0, \\ r_t + D\, r_x = \eta_0\, M\, r, \\ r_+(t, 0) = u_+(t), \\ r_-(t, L) = R_-^d \end{cases} \tag{4.2}$$

has a solution $r \in C([0, T], L^2((0, L); \mathbb{R}^2))$. Moreover, for the boundary traces of the solution, we have $r_+(\cdot, L), r_-(\cdot, 0) \in L^2(0, T)$.

For $x = (x_+, x_-) \in \mathbb{R}^2$, we use the notation $\|x\|_{\mathbb{R}^2} = \sqrt{x_+^2 + x_-^2}$. For $u = (u_+, u_-) \in (L^2(0, T))^2$ and $R = (R_+, R_-) \in (L^2(0, T))^2$, define the objective function

$$J(u, R)$$

$$= \int_0^T \tfrac{1}{2}\, (u_+(t))^2\, dt + \gamma \int_{t_0}^{T - t_0} \|(R_+(t) - R_+^d,\ R_-(t) - R_-^d\|_{\mathbb{R}^2}\, dt. \tag{4.3}$$

Then, if L is sufficiently small and T and $t_0 < T$ are sufficiently large, the system is nodal profile exactly controllable and Theorem 3 is applicable for the optimal control problem

$$\begin{cases} \min_{u_+ \in L^2(0, T)} J(u_+, (r_+(\cdot, L), r_-(\cdot, L))) \\ \text{subject to (4.2).} \end{cases} \tag{4.4}$$

In fact, the result of Theorem 3 can be interpreted as a *finite-time turnpike result* (or exact turnpike), where the system is driven to a desired stationary state in finite time.

5 Conclusion

We have shown that a finite-time turnpike phenomenon occurs for problems of optimal control with non-differentiable norm tracking terms.

We have first considered systems that are governed by ordinary differential equations. In the objective functions, L^1-norm tracking terms are used. The finite-time turnpike means that after finite time, the optimal state reaches the desired state. For infinite-dimensional systems, we have shown that a finite-time turnpike phenomenon occurs for problems of optimal control for systems that are exactly controllable with a max-norm-type tracking term and a weighted L^1-norm tracking term. For systems that are nodal profile exactly controllable, we have shown that a finite-time turnpike phenomenon occurs with an L^2-norm tracking term.

Acknowledgments This project has received funding from the European Research Council (ERC) under the European Union's Horizon 2020 research and innovation programme (grant agreement No. 694126-DyCon). The work of the third author has been funded by the Alexander von Humboldt-Professorship program, the European Union's Horizon 2020 research and innovation programme under the Marie Sklodowska-Curie grant agreement No.765579-ConFlex, grant MTM2017-92996-C2-1-R COSNET of MINECO (Spain), ELKARTEK project KK-2018/00083 ROAD2DC of the Basque Government, ICON of the French ANR and Nonlocal PDEs: Analysis, Control and Beyond, AFOSR Grant FA9550-18-1-0242, and Transregio 154, Mathematical Modelling, Simulation and Optimization using the Example of Gas Networks, of the German DFG, project C08.

References

1. Damm, T., Grüne, L., Stieler, M., Worthmann, K.: An exponential turnpike theorem for dissipative discrete time optimal control problems. SIAM J. Control Optim. **52**, 1935–1957 (2014)
2. Dorfman, R., Samuelson, P.A., Solow, R.M.: Linear Programming and Economic Analysis. McGraw-Hill, New York (1958)
3. Faulwasser, T., Bonvin, D.: On the design of economic NMPC based on an exact turnpike property. IFAC-PapersOnLine **48**, 525–530 (2015)
4. Grüne, L., Guglielmi, R.: Turnpike properties and strict dissipativity for discrete time linear quadratic optimal control problems. SIAM J. Control Optim. **56**, 1282–1302 (2018)
5. Grüne, L., Schaller, M.: Exponential sensitivity and turnpike analysis for linear quadratic optimal control of general evolution equations. J. Differ. Equ. **268**, 7311–7341 (2020)
6. Gugat, M.: A turnpike result for convex hyperbolic optimal boundary control problems. Pure Appl. Funct. Anal. **4**, 849–866 (2019)
7. Gugat, M., Hante, F.: On the turnpike phenomenon for optimal boundary control problems with hyperbolic systems. SIAM J. Control Optim. **57**, 264–289 (2019)
8. Gugat, M., Zuazua, E.: Exact penalization of terminal constraints for optimal control problems. Optim. Control Appl. Meth. **37**, 1329–1354 (2016)
9. Gugat, M., Herty, M., Sacher, V.: Flow control in gas networks: exact controllability to a given demand. Math. Methods Appl. Sci. **34**, 745–757 (2011)
10. Gugat, M., Trélat, E., Zuazua, E.: Optimal Neumann control for the 1D wave equation: finite horizon, infinite horizon, boundary tracking terms and the turnpike property. Syst. Control Lett. **90**, 61–70 (2016)
11. Li, T.-T., Wang, K., Gu, Q.: Exact Boundary Controllability of Nodal Profile for Quasilinear Hyperbolic Systems. SpringerBriefs in Mathematics. Springer, Singapore (2016)
12. Phillips, R.S.: A note on the abstract Cauchy problem. Proc. Nat. Acad. Sci. U.S.A. **40**, 244–248 (1954)
13. Porretta, A., Zuazua, E.: Long time versus steady state optimal control. SIAM J. Control Optim. **51**, 4242–4273 (2013)
14. Trelat, E., Zhang, C.: Integral and measure-turnpike properties for infinite-dimensional optimal control systems. Math. Control Signals Syst. **30**, 3 (2018)
15. Trelat, E., Zhang, C., Zuazua, E.: Steady-state and periodic exponential turnpike property for optimal control problems in Hilbert spaces. SIAM J. Control Optim. **56**, 1222–1252 (2018)
16. Tucsnak M., Weiss, G.: Observation and Control for Operator Semigroups. Birkhäuser Advanced Texts, Basel (2009)

17. Zaslavski, A.: Existence and structure of optimal solutions of infinite dimensional control problems. Appl. Math. Opt. **42**, 291–313 (2000)
18. Zaslavski, A.: Turnpike Properties in the Calculus of Variations and Optimal Control. Springer, New York (2006)
19. Zaslavski, A.: Turnpike Conditions in Infinite Dimensional Optimal Control. Springer Nature, Cham (2019)

On the Eigenvalue Distribution for a Beam with Attached Masses

Julia Kalosha, Alexander Zuyev, and Peter Benner

Abstract We study a mathematical model of a hinged flexible beam with piezoelectric actuators and electromagnetic shaker in this chapter. The shaker is modeled as a mass and spring system attached to the beam. To analyze free vibrations of this mechanical system, we consider the corresponding spectral problem for a fourth-order differential operator with interface conditions that characterize the shaker dynamics. The characteristic equation is studied analytically, and asymptotic estimates of eigenvalues are obtained. The eigenvalue distribution is also illustrated by numerical simulations under a realistic choice of mechanical parameters.

1 Introduction

The stabilization problem for distributed parameter mechanical systems with flexible beams has been the subject of investigations of many authors. Without any pretense to represent an exhaustive list, we just mention some of the most related works. In the monograph [1], the problem of strong asymptotic stabilization of vibration systems with Euler–Bernoulli beams has been investigated. In [2], spectral methods for the investigation of strong stability of distributed parameter systems in

J. Kalosha
Institute of Applied Mathematics and Mechanics, National Academy of Sciences of Ukraine, Sloviansk, Ukraine

A. Zuyev (✉)
Max Planck Institute for Dynamics of Complex Technical Systems, Magdeburg, Germany

Institute of Applied Mathematics and Mechanics, National Academy of Sciences of Ukraine, Sloviansk, Ukraine

Donbas State Pedagogical University, Sloviansk, Ukraine
e-mail: zuyev@mpi-magdeburg.mpg.de

P. Benner
Max Planck Institute for Dynamics of Complex Technical Systems, Magdeburg, Germany
e-mail: benner@mpi-magdeburg.mpg.de

G. Sklyar, A. Zuyev (eds.), *Stabilization of Distributed Parameter Systems: Design Methods and Applications*, SEMA SIMAI Springer Series 2,
https://doi.org/10.1007/978-3-030-61742-4_3

Hilbert spaces are presented on a common basis, and some examples of controlled beams are analyzed. The stabilization of an essentially nonlinear system consisting of a beam attached to the disc with torque control is achieved in [3].

The dynamic behavior of a non-uniform cantilever Euler–Bernoulli beam with boundary control is investigated in the paper [4]. It is emphasized that asymptotic properties of the eigenvalues of the corresponding infinitesimal generator play an important role for proving the exponential stability of the closed-loop system. A numerical scheme that preserves the exponential stability of the original continuous Euler–Bernoulli model with a clamped end and boundary control is proposed in [5] without introducing additional numerical viscosity. The Euler–Bernoulli beam with one clamped end is considered in [6] as a control system with two-dimensional input and output. It is shown that the spectrum of the generator is located in the open upper half-plane for the considered model, and asymptotic approximations of the eigenvalues are proposed. In the paper [7], the general eigenvalue problem is formulated for the Timoshenko and the Euler–Bernoulli beam models under different boundary conditions. A regularization process for the characteristic equation is discussed, and numerically stable forms of the original hyperbolic expressions are obtained.

Over the last few years, stability and stabilization problems for laminated beams and beam networks have received a lot of attention due to their rich theoretical context and engineering applications (see, e.g., [8–11] and references therein). Stability conditions for elastic structures formed by serially connected flexible beams with different types of boundary conditions are obtained and illustrated by numerical examples in [12, 13]. In the paper [14], the covariance functions are computed for the transverse displacement of the Euler–Bernoulli beam under stochastic excitation. The Euler–Bernoulli model with one clamped end and another simply supported end under the action of additive white noise is studied in [15]. It is shown that there exists a global random attractor for the considered stochastic system, and the Hausdorff dimension of the global random attractors is estimated.

In contrast to the above-mentioned publications, in this work we focus on the mathematical model of a simply supported beam of length l with k piezoelectric actuators and an electromagnetic shaker described in [16]. This mathematical model is represented by the following abstract differential equation [16]:

$$\frac{d}{dt}\,\xi(t) = \mathcal{A}\xi(t) + \mathcal{B}y, \quad \xi(t) \in X, \; y \in \mathbb{R}^{k+1}. \tag{1}$$

Here,

$$X = \{\xi = (u, v, p, q)^T \mid u \in \overset{\circ}{H}{}^2(0,l), \; v \in L^2(0,l), \; p, q \in \mathbb{C}\}$$

is the Hilbert space with the inner product

$$\left\langle \begin{pmatrix} u_1 \\ v_1 \\ p_1 \\ q_1 \end{pmatrix}, \begin{pmatrix} u_2 \\ v_2 \\ p_2 \\ q_2 \end{pmatrix} \right\rangle_X = \int_0^l \left(E(x)I(x)u_1''(x)\bar{u}_2''(x) + \rho(x)v_1(x)\bar{v}_2(x)\right) dx$$

$$+ \varkappa p_1 \bar{p}_2 + m q_1 \bar{q}_2,$$

and $y = (F, M_1, \dots, M_k)^T$ is considered as the control. The components of y have the following physical meaning [16]: F is the control force applied by the shaker to the beam at a point $x = l_0$, $l_0 \in (0, l)$, and M_j characterizes the action of the j-th piezoelectric actuator, $j = 1, 2, \dots, k$. Here, $\rho(x) > 0$, $E(x)I(x) \in C^2[0, l]$, $E(x)I(x) > 0$ for all $x \in (0, l)$, m and $\varkappa$ are assumed to be positive constants. Throughout the text, the prime stands for the derivative with respect to the spatial variable x. The Sobolev spaces $\overset{\circ}{H}{}^k(0, l)$ consist of all functions $u \in H^k(0, l)$ such that $u(0) = u(l) = 0$.

The linear differential operator $\mathcal{A}$ with the domain

$$\mathcal{D}(\mathcal{A}) = \left\{ \xi = \begin{pmatrix} u \\ v \\ p \\ q \end{pmatrix} \in X : \begin{array}{l} u \in H^4(0, l_0) \cap H^4(l_0, l), \\ u''(0) = u''(l) = 0, \\ u''(l_0 - 0) = u''(l_0 + 0), \\ v \in \overset{\circ}{H}{}^2(0, l), \\ p = u(l_0), \quad q = v(l_0) \end{array} \right\}$$

is defined by the following rule:

$$\mathcal{A} : \xi = \begin{pmatrix} u \\ v \\ p \\ q \end{pmatrix} \mapsto \mathcal{A}\xi = \begin{pmatrix} v \\ -\frac{1}{\rho}(EIu'')'' \\ q \\ -\frac{1}{m}\left\{ \varkappa p - (EIu'')'\Big|_{l_0 - 0} + (EIu'')'\Big|_{l_0 + 0} \right\} \end{pmatrix}. \quad (2)$$

The linear operator $\mathcal{B} : \mathbb{R}^{k+1} \to X$ is given by its matrix

$$\mathcal{B} = \begin{pmatrix} 0 & 0 & \dots & 0 \\ 0 & \frac{1}{\rho}\psi_1'' & \dots & \frac{1}{\rho}\psi_k'' \\ 0 & 0 & \dots & 0 \\ \frac{1}{m} & 0 & \dots & 0 \end{pmatrix}, \quad (3)$$

where $\psi_j \in C^2(0,l)$ is the shape function of the j-th piezoelectric actuator such that $\operatorname{supp}\psi_j \cap \{0, l_0, l\} = \emptyset$, $j = 1, 2, \ldots, k$. A state feedback law $y = \mathcal{K}\xi$ has been proposed in [16] such that $\mathcal{K}: X \to \mathbb{R}^{k+1}$ is a bounded linear operator,

$$\xi = \begin{pmatrix} u \\ v \\ p \\ q \end{pmatrix} \in X \mapsto \mathcal{K}\xi = -\begin{pmatrix} \alpha_0 q \\ \alpha_1 \int_0^l \psi_1''(x)v(x)dx \\ \vdots \\ \alpha_k \int_0^l \psi_k''(x)v(x)dx \end{pmatrix} \in \mathbb{R}^{k+1}, \tag{4}$$

and $\alpha_0, \alpha_1, \ldots, \alpha_k$ are positive tuning parameters. We refer to the following stability result for the closed-loop system (1) with $y = \mathcal{K}\xi(t)$.

Theorem 1 ([16]) *Let $\tilde{\mathcal{A}} = \mathcal{A} + \mathcal{B}\mathcal{K}$, $\tilde{\mathcal{A}} : \mathcal{D}(\mathcal{A}) \to X$, where the operators $\mathcal{A}$, $\mathcal{B}$, and $\mathcal{K}$ are given by* (2)–(4)*, and let $\alpha_j > 0$ for $j = 0, 1, \ldots, k$. Then, the abstract Cauchy problem*

$$\frac{d}{dt}\xi(t) = \tilde{\mathcal{A}}\xi(t), \tag{5}$$

$$\xi(0) = \xi_0 \in X \tag{6}$$

is well-posed on $t \geq 0$, and the solution $\xi = 0$ of the closed-loop system (5) *is stable in the sense of Lyapunov.*

Note that the above theorem does not guarantee asymptotic stability of the equilibrium $\xi = 0$ as the limit behavior of trajectories is determined by invariant subsets of the set $M = \{\xi \in \mathcal{D}(\mathcal{A}) \mid \dot{E}(\xi) = 0\}$, where $E(\xi)$ is the weak Lyapunov functional constructed in [16]. As it is well-known, the analysis of such invariant subsets is related to spectral properties of the infinitesimal generator (see, e.g., [17] and references therein).

Moreover, the distribution of eigenvalues is one of the crucial characteristics of distributed parameter systems related to their controllability, observability, and stabilizability properties because of the Ingham-type theorems [18] with important applications to one-dimensional wave-like equations [19, 20]. Thus, the goal of this chapter is to study the spectral problem for $\mathcal{A}$ and analyze the distribution of roots of the corresponding frequency equation. In Sect. 2, we derive an approximate frequency equation and prove its equivalence to the exact one in the sense of limit behavior of the roots. An estimate of the growth rate of eigenfrequencies is obtained in Sect. 3. We will present the results of numerical simulations to illustrate the behavior of the roots of both equations. For the sake of simplicity, we will assume E and I to be positive constants in the sequel.

2 Spectral Problem

In order to obtain the characteristic equation, we consider the spectral problem

$$\mathcal{A}\xi = \lambda\xi, \quad \xi \in \mathcal{D}(\mathcal{A}), \tag{7}$$

where $\lambda = i\omega$. This problem reduces to the following set of equations with respect to the components of ξ:

$$\begin{aligned} v &= \lambda u, \\ \frac{EI}{\rho}\frac{d^4u}{dx^4} + \lambda v &= 0, \\ q &= \lambda p, \\ L - \varkappa p &= \lambda m q, \end{aligned} \tag{8}$$

and the condition $\xi \in \mathcal{D}(\mathcal{A})$ yields

$$\begin{aligned} &u(0) = u(l) = 0, \\ &u''(0) = u''(l) = 0, \qquad u^{(j)}(l_0 - 0) - u^{(j)}(l_0 + 0) = 0, \quad j = \overline{0, 2}, \\ &v(0) = v(l) = 0, \qquad EI(u'''(l_0 - 0) - u'''(l_0 + 0)) = L, \\ &p = u(l_0), \quad q = v(l_0). \end{aligned} \tag{9}$$

Let us first consider the second equation in (8) with respect to $u(x)$:

$$\frac{d^4u}{dx^4} - \mu^4 u = 0, \quad x \neq l_0, \tag{10}$$

where $\mu = \left(\frac{\rho}{EI}\omega^2\right)^{1/4}$ is treated as the new spectral parameter of problem (7) and (9). The general solution of (10) can be represented in the intervals of continuity of $u(x)$ as

$$u(x) = C_1 e^{-\mu x} + C_2 e^{\mu x} + C_3 \sin \mu x + C_4 \cos \mu x, \quad x \neq l_0.$$

Let us denote

$$U(x) = \begin{pmatrix} u_0(x) \\ u_1(x) \\ u_2(x) \\ u_3(x) \end{pmatrix} = \begin{pmatrix} u(x) \\ u'(x) \\ u''(x) \\ u'''(x) \end{pmatrix},$$

and then the differential equation that defines the eigenfunctions of (7) and (9) can be written for $x \in (0, l_0) \cup (l_0, l)$ as

$$\frac{d}{dx}U(x) = MU(x), \quad M = \begin{pmatrix} 0 & 1 & 0 & 0 \\ 0 & 0 & 1 & 0 \\ 0 & 0 & 0 & 1 \\ \mu^4 & 0 & 0 & 0 \end{pmatrix}. \tag{11}$$

The general solution of (11) is represented by the matrix exponential as follows:

$$U(x) = \begin{cases} e^{xM}U(0), & x \in (0, l_0), \\ e^{(x-l)M}U(l), & x \in (l_0, l), \end{cases} \tag{12}$$

where

$$e^{xM} = \begin{pmatrix} z_1(x) & z_2(x) & z_3(x) & z_4(x) \\ \mu^4 z_4(x) & z_1(x) & z_2(x) & z_3(x) \\ \mu^4 z_3(x) & \mu^4 z_4(x) & z_1(x) & z_2(x) \\ \mu^4 z_2(x) & \mu^4 z_3(x) & \mu^4 z_4(x) & z_1(x) \end{pmatrix},$$

$$z_1(x) = \frac{1}{2}(\cosh \mu x + \cos \mu x), \quad z_2(x) = \frac{1}{2\mu}(\sinh \mu x + \sin \mu x),$$

$$z_3(x) = \frac{1}{2\mu^2}(\cosh \mu x - \cos \mu x), \quad z_4(x) = \frac{1}{2\mu^3}(\sinh \mu x - \sin \mu x).$$

The constants of integration $u_0(0)$, $u_2(0)$, $u_0(l)$, and $u_2(l)$ are equal to zero due to the boundary conditions of the spectral problem, while the other four can be obtained from the interface conditions at the point $x = l_0$. These conditions may be represented as the system of linear algebraic equations $\mathcal{M}(u_1(0), u_3(0), u_1(l), u_3(l))^T = 0$, whose matrix is

$$\mathcal{M} = \begin{pmatrix} z_2(l_0) & z_4(l_0) & -z_2(l_0 - l) & -z_4(l_0 - l) \\ z_1(l_0) & z_3(l_0) & -z_1(l_0 - l) & -z_3(l_0 - l) \\ \mu^4 z_4(l_0) & z_2(l_0) & -\mu^4 z_4(l_0 - l) & -z_2(l_0 - l) \\ \mu^4 z_3(l_0) - \hat{\mu} z_2(l_0) & z_1(l_0) - \hat{\mu} z_4(l_0) & -\mu^4 z_3(l_0 - l) & -z_1(l_0 - l) \end{pmatrix},$$

where $\hat{\mu} = \frac{\varkappa}{EI} - \frac{m}{\rho}\mu^4$. The last equality can be obtained from the conditions $\varkappa p + \lambda m q = EI(u_3(l_0 - 0) - u_3(l_0 + 0))$, $q = \lambda p$. The value of p is treated as the limit of $u_0(x)$ at $x = l_0$. Thus, we obtain the following frequency equation:

$$\det \mathcal{M} = 0, \tag{13}$$

where

$$\det \mathcal{M} = \frac{m}{4\mu\rho}\{(\cosh \mu(l-2l_0) - \cosh \mu l)\sin \mu l + (\cos \mu(l-2l_0) - \cos \mu l)\sinh \mu l\}$$
$$- \frac{\sin \mu l \sinh \mu l}{\mu^2} + \frac{\varkappa}{4EI\mu^5}\{(\cosh \mu l - \cosh \mu(l-2l_0))\sin \mu l$$
$$-(\cos \mu(l-2l_0) + \cos \mu l)\sinh \mu l\}.$$

If μ satisfies the above equation, then rank $\mathcal{M} \leqslant 3$ and it is possible to choose a nonzero set of values $u_1(0)$, $u_3(0)$, $u_1(l)$, and $u_3(l)$. Let us take, for instance, $u_3(l) = 1$, and then the other boundary values can be found from the following system of linear algebraic equations:

$$\mathcal{M}_3(u_1(0), u_3(0), u_1(l))^T = (z_4(l_0 - l), z_3(l_0 - l), z_2(l_0 - l))^T, \tag{14}$$

where

$$\mathcal{M}_3 = \begin{pmatrix} z_2(l_0) & z_4(l_0) & -z_2(l_0 - l) \\ z_1(l_0) & z_3(l_0) & -z_1(l_0 - l) \\ \mu^4 z_4(l_0) & z_2(l_0) & -\mu^4 z_4(l_0 - l) \end{pmatrix}.$$

Thus, solution (12) of Eq. (11) contains the function $u(x)$ of the following form:

$$u(x) = \begin{cases} z_2(x)u_1(0) + z_4(x)u_3(0), & x \in (0, l_0), \\ z_2(x-l)u_1(l) + z_4(x-l), & x \in (l_0, l), \end{cases}$$

and the set of boundary values is determined uniquely by system (14), provided that

$$\det \mathcal{M}_3 = \frac{\sinh \mu l_0 \sin \mu l + \sin \mu l_0 \sinh \mu l}{2\mu^2} \neq 0. \tag{15}$$

Then, the rest of the components of ξ are determined by system (8). This completes the procedure of solving the spectral problem (7).

3 Frequency Analysis

The determinant of $\mathcal{M}$ in (13) admits the following asymptotic representation for $\mu \to +\infty$:

$$\det \mathcal{M} = \frac{me^{\mu l}}{8\rho\mu}\left(\Phi_0(\mu) + o(1)\right),$$

where

$$\Phi_0(\mu) = 2\sin\mu(l - l_0)\sin\mu l_0 - \sin\mu l.$$

Thus, Eq. (13) is equivalent to the following one if $\mu \neq 0$:

$$\Phi_0(\mu) + \Phi_1(\mu) = 0, \tag{16}$$

where

$$\Phi_1(\mu) = e^{-\mu l}\{2\sinh\mu l\cos\mu(l - 2l_0) - 2\sinh\mu l\cos\mu l - 2\cosh\mu l\sin\mu l$$
$$+2\sin\mu l\cosh\mu(l - 2l_0) + e^{\mu l}(\cos\mu l + \sin\mu l - \cos\mu(l - 2l_0)) - \frac{8}{m\mu}\rho\sinh\mu l\sin\mu l\}$$
$$+\frac{2\varkappa\rho e^{-\mu l}}{EIm\mu^4}\{(\cosh\mu l - \cosh\mu(l - 2l_0))\sin\mu l + (\cos\mu l - \cos\mu(l - 2l_0))\sinh\mu l\},$$

and $\Phi_1(\mu) \to 0$ as $\mu \to +\infty$. Note that the distribution of the roots of $\Phi_0(\mu)$ is determined by the length parameters l and l_0 only. We will show below that the roots of (16) can be approximated by the roots of

$$\Phi_0(\mu) = 0, \tag{17}$$

provided that $\mu > 0$ is large enough.

Proposition 1 *Assume that the number l_0/l is rational. Let $\bar{\mu}_1 < \bar{\mu}_2 < \ldots$ be the positive roots of Eq.* (17)*. Then, for every $\varepsilon > 0$, there is an $M > 0$ such that for every $\bar{\mu}_j > M$, there exists a unique root $\mu_j \in I_j = (\bar{\mu}_j - \varepsilon; \bar{\mu}_j + \varepsilon)$ of* (16)*.*

It means that every root of Eq. (16) is located in an ε-neighborhood of the corresponding root of Eq. (17), and this neighborhood does not contain any other root of (16). Besides, it is important to ensure that there are no "large" roots of Eq. (16) outside the ε-neighborhoods mentioned above, as stated in the next proposition.

Proposition 2 *Let the assumptions of Proposition 1 be satisfied, and let*

$$S = (M; +\infty) \setminus \bigcup_{j=1}^{\infty}(\mu_j - \varepsilon; \mu_j + \varepsilon).$$

Then, $\Phi_0(\mu) + \Phi_1(\mu) \neq 0$ for all $\mu \in S$.

Proof of Propositions 1 and 2 In case when the numbers l_0 and l are commensurable, the function $\Phi_0(\mu)$ is periodic and its zeros $0 < \bar{\mu}_1 < \bar{\mu}_2 < \ldots$ are simple, $\Phi_0'(\mu_j) \neq 0$. Without loss of generality, we assume that a number $\varepsilon > 0$ is chosen small enough so that $\varepsilon < (\bar{\mu}_{j+1} - \bar{\mu}_j)/2$ (the intervals I_j and I_{j+1} are disjoint) for all $j \geq 1$, and

$$\inf_{|\mu-\bar{\mu}_j|\leq\varepsilon,\, j\geq 1} |\Phi_0'(\mu)| \geqslant K \tag{18}$$

with some constant $K > 0$. By construction, all positive roots of the equation $\Phi_0(\mu) = 0$ are contained in the open set $I = \cup_{j\geq 1} I_j$. Hence, $\Phi_0(\mu) \neq 0$ at each $\mu \in S_0 = [0, +\infty) \setminus I$. By exploiting the periodicity and continuity of Φ_0, we conclude from the Weierstrass extreme value theorem that there exists a $\delta > 0$ such that

$$|\Phi_0(\mu)| \geq \delta \quad \text{for all } \mu \in S_0. \tag{19}$$

From the mean value theorem, we know that

$$\Phi_0(\mu) = \Phi_0'(\chi)(\mu - \bar{\mu}_j), \tag{20}$$

where $\chi = \bar{\mu}_j + \Theta(\mu - \bar{\mu}_j)$, $\Theta \in (0, 1)$. Then, (18) and (20) imply that

$$|\Phi_0(\mu_j \pm \varepsilon)| \geq K\varepsilon \text{ and } \Phi_0(\mu_j - \varepsilon)\Phi_0(\mu_j + \varepsilon) < 0 \quad \text{for all } j = 1, 2, \dots . \tag{21}$$

Since $\Phi_1(\mu) \to 0$ and $\Phi_1'(\mu) \to 0$ as $\mu \to +\infty$, we take an $M > 0$ such that

$$|\Phi_1(\mu)| < \min\{K\varepsilon, \delta\} \text{ and } |\Phi_1'(\mu)| < K/2 \quad \text{for all } \mu > M - \varepsilon. \tag{22}$$

From (21) and (22), it follows that the continuous function $\Phi(\mu) = \Phi_0(\mu) + \Phi_1(\mu)$ has values of opposite sign at $\bar{\mu}_j \pm \varepsilon$, provided that $\bar{\mu}_j > M$. Then, there exists a $\mu_j \in I_j = (\bar{\mu}_j - \varepsilon; \bar{\mu}_j + \varepsilon)$ such that $\Phi(\mu_j) = 0$ by the intermediate value theorem.

The uniqueness of the root μ_j in I_j can be proved by contradiction. Let $\mu_j^* \in I_j$ be such that

$$\Phi_0(\mu_j^*) = -\Phi_1(\mu_j^*), \quad \mu_j^* \neq \mu_j. \tag{23}$$

On the one hand, the integral representations

$$\Phi_0(\mu_j^*) = \Phi_0(\mu_j) + \int_{\mu_j}^{\mu^*} \Phi_0'(\zeta)d\zeta, \quad \Phi_1(\mu_j^*) = \Phi_1(\mu_j) + \int_{\mu_j}^{\mu^*} \Phi_1'(\zeta)d\zeta$$

together with the condition $\Phi_0(\mu_j) = -\Phi_1(\mu_j)$ yield

$$\int_{\mu_j}^{\mu^*} (\Phi_0'(\zeta) + \Phi_1'(\zeta))d\zeta = 0. \tag{24}$$

On the other hand, inequalities (18) and (22) imply

$$\left| \int_{\mu_j}^{\mu^*} (\Phi_0'(\zeta) + \Phi_1'(\zeta))d\zeta \right| \geq |\mu_j - \mu^*| \inf_{\zeta \in I_j} |\Phi_0'(\zeta) + \Phi_1'(\zeta)| \geq$$

$$\geq |\mu_j - \mu^*| \big(\inf_{\zeta \in I_j} |\Phi_0'(\zeta)| - \sup_{\zeta \in I_j} |\Phi_1'(\zeta)| \big) \geq \frac{K|\mu_j - \mu^*|}{2} > 0.$$

The above inequality contradicts (24); hence, μ_j is the only zero of $\Phi(\mu)$ in I_j, which proves Proposition 1.

By taking into account (19) and (22), we conclude that $|\Phi_0(\mu)| > |\Phi_1(\mu)|$ whenever $\mu \in S_0$ and $\mu > M$. The latter implies the assertion of Proposition 2. □

4 Numerical Simulation Results

Let $0 < \mu_1 \leq \mu_2 \leq \ldots \leq \mu_N$ be the first N positive roots of Eq. (13). Then, the corresponding eigenvalues λ_j of (7) and the modal frequencies ν_j in Hz are expressed as

$$\lambda_j = i\sqrt{\frac{EI}{\rho}}\mu_j^2, \quad \nu_j = \frac{\sqrt{EI}}{2\pi\sqrt{\rho}}\mu_j^2, \quad j = 1, 2, \ldots, N.$$

To test the approximation of μ_j by solutions of the truncated equation (17), we also compute numerically the roots $\bar{\mu}_j$ of the transcendent equation $\Phi_0(\bar{\mu}) = 0$ and the corresponding frequency parameters $\bar{\nu}_j = \frac{\sqrt{EI}}{2\pi\sqrt{\rho}}\bar{\mu}_j^2$ for $N = 22$. These numerical results are summarized in Table 1 for the following realistic values of mechanical parameters:

$$l = 1.905\,\text{m}, \; l_0 = 1.4\,\text{m}, \; \rho_0 = 2700\,\text{kg/m}^3, \; S = 2.25 \cdot 10^{-4}\text{m}^2, \; \rho = \rho_0 S,$$

$$E = 6.9 \cdot 10^{10}\,\text{Pa}, \; I = 1.6875 \cdot 10^{-10}\,\text{m}^4, \; m = 0.1\,\text{kg}, \; \varkappa = 7\,\text{N/mm}. \tag{25}$$

The above mechanical parameters correspond to the experimental setup described in [21].

Table 1 Modal frequencies under the choice of parameters (25)

j	$\bar{\mu}_j$	μ_j	$\bar{\nu}_j$	ν_j
1	2.616	2.552	4.767	4.537
2	4.714	4.573	15.485	14.570
3		5.618		21.994
4	6.553	6.608	29.921	30.427
5	7.460	8.198	38.774	46.830
6	9.309	9.721	60.378	65.850
7	11.407	11.494	90.661	92.061
8	13.013	13.142	117.997	120.342
9	14.018	14.501	136.914	146.510
10	16.009	16.226	178.565	183.445
11	18.085	18.113	227.881	228.610
12	20.648	20.988	297.075	306.918
13	22.711	22.846	359.376	363.683
14	24.732	24.734	426.184	426.273
15	25.803	26.084	463.925	474.083
16	27.318	27.554	519.992	528.997
17	29.411	29.497	602.735	606.252
18	31.318	31.325	683.413	683.723
19	32.238	32.533	724.138	737.484
20	34.007	34.172	805.799	813.665
21	36.107	36.158	908.419	910.975
22	37.814	37.863	996.296	998.922

Note that the truncated frequency Eq. (17) has an additional positive root $\bar{\mu}_0 \approx 0.9949$ that does not correspond to any root of Eq. (13), and there is the root μ_3 of (13), while Eq. (17) does not have any root in the corresponding interval. These features are observed in low frequency range. But for large values of μ, the agreement between solutions of the frequency Eqs. (13) and (17) is quite good, as seen in Table 1 and predicted by Propositions 1 and 2.

In the particular case $l = 2l_0$, it is easy to write the positive roots of the transcendent equation (17) explicitly:

$$\bar{\mu}_j = \frac{\pi}{l}\left(\left\{\frac{j}{2}\right\} + 2\left[\frac{j}{2}\right]\right), \quad j = 1, 2, \ldots. \tag{26}$$

The above formula together with Propositions 1 and 2 implies that the spectral parameters μ_j grow linearly with j, i.e., the modal frequencies ν_j and the eigenvalues λ_j of (7) grow quadratically with j for large j in the considered case. The growth rate of μ_j and $\bar{\mu}_j$ in the case of mechanical parameters (25) is shown in Fig. 1.

To illustrate the behavior of the corresponding eigenvectors $\xi^1, \xi^2, \ldots, \xi^N$ of (7), we plot the graphs of their u-component in Figs. 2 and 3. Note that condition (15) holds for each $\mu = \mu_j$ in the considered case; thus, each ξ^j is uniquely defined up to normalization. As the functions $u^j(x)$ describe the transverse displacement of the beam corresponding to the spectral parameter μ_j, we will refer to $u^j(x)$ as the j-th eigenmode of vibration. The presented eigenmodes correspond to the choice of $l_0 = 1.4$ m in Fig. 2 and $l_0 = l/2$ in Fig. 3, respectively, for $N = 4$. In both figures, the functions $u^j(x)$ are normalized in the sense of L^2-norm on $[0, l]$.

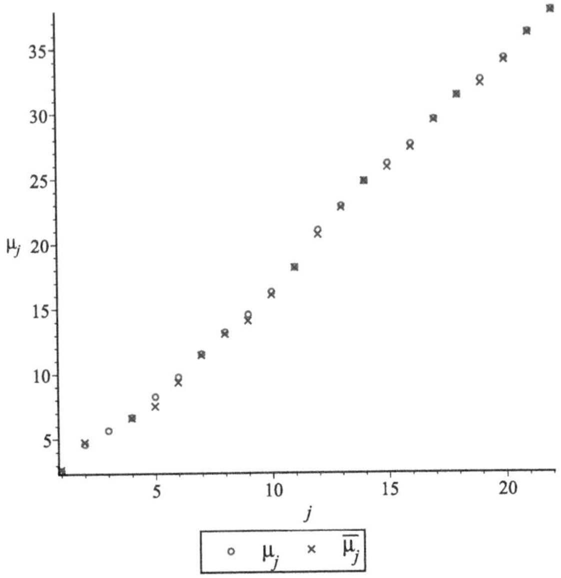

Fig. 1 Spectral parameters μ_j and $\bar{\mu}_j$

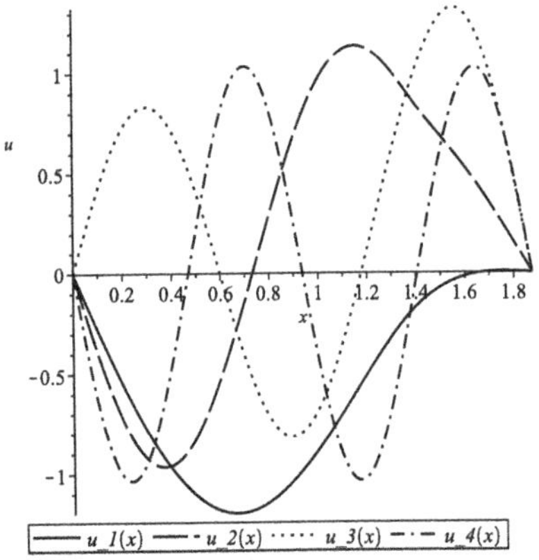

Fig. 2 Eigenmodes for $l_0 = 1.4$ m

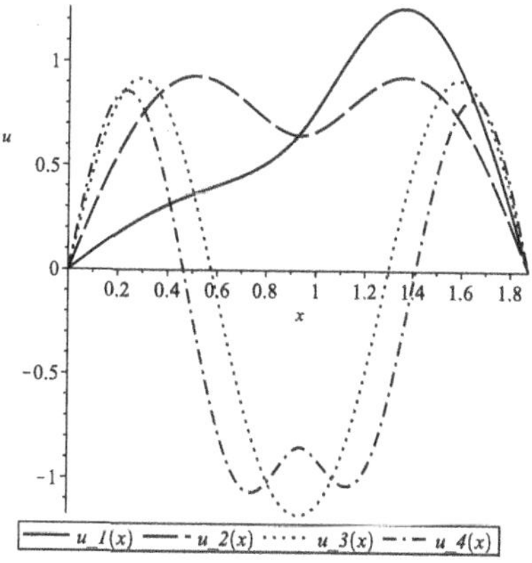

Fig. 3 Eigenmodes for $l_0 = l/2$

5 Conclusion

The approximation results of Sect. 3 together with the linear growth condition of the form (26) generalize known asymptotic properties of eigenvalues of the standard Euler–Bernoulli beam (cf. [1, Chapter 4]) to the model with attached masses. In future work, we expect to apply Propositions 1 and 2 for the analysis of the limit behavior of trajectories to characterize attractors of the infinite-dimensional closed-loop system (5) for different values of tuning parameters of the feedback law (4). The results obtained in this chapter are planned to be extended to other classes of elastic structures, particularly, to the rotating Timoshenko beam with attached masses [22] and beam systems with passive joints [23].

References

1. Luo, Z.-H., Guo, B.-Z., Morgul O.: Stability and Stabilization of Infinite Dimensional Systems with Applications. Springer, London (1999)
2. Oostveen, J.: Strongly Stabilizable Distributed Parameter Systems. SIAM, Philadelphia (2000)
3. Coron, J.-M.: Control and Nonlinearity. AMS, Providence (2007)
4. Guo, B.-Z.: Riesz basis property and exponential stability of controlled Euler–Bernoulli beam equations with variable coefficients. SIAM J. Control Optim. **40**, 1905–1923 (2002)
5. Liu, J., Guo, B.-Z.: A novel semi-discrete scheme preserving uniformly exponential stability for an Euler–Bernoulli beam. Syst. Control Lett. **134**, 1–18 (2019) https://doi.org/10.1016/j.sysconle.2019.104518
6. Shubov, M.A.: Location of eigenmodes of Euler—Bernoulli beam model under fully non-dissipative boundary conditions. Procee. R. Soc. A **475**, 1–20 (2019) https://doi.org/10.1098/rspa.2019.0544
7. Khasawneh, F.A., Segalman, D.: Exact and numerically stable expressions for Euler–Bernoulli and Timoshenko beam modes. Appl. Acoust. **151**, 215–228 (2019)

8. Liu, W., Luan, Y., Liu, Y., Li, G.: Well-posedness and asymptotic stability to a laminated beam in thermoelasticity of type III. Math. Methods Appl. Sci. **43**, 3148–3166 (2019). https://doi.org/10.1002/mma.6108
9. Apalara, T.A., Raposo, C.A., Nonato, C.A.S.: Exponential stability for laminated beams with a frictional damping. Archiv der Math. **114**, 471–480 (2020). https://doi.org/10.1007/s00013-019-01427-1
10. Apalara, T.A.: On the stability of a thermoelastic laminated beam. Acta Math. Sci. **39**, 1517–1524 (2019)
11. Mustafa, M.I.: On the stabilization of viscoelastic laminated beams with interfacial slip. Z. Angew. Math. Phys. **69**. (2018) https://doi.org/10.1007/s00033-018-0928-7
12. Chen, G., Delfour, M., Krall, A., Payre, G.:. Modeling, stabilization and control of serially connected beams. SIAM J. Control Optim. **25**, 526–546 (1987)
13. Lagnese, J.E., Leugering, G., Schmidt, E.J.P.G.: Modeling, Analysis and Control of Dynamic Elastic Multi-Link Structures. Birkhäuser, Boston (1994)
14. Śniady, P., Podworna, M., Idzikowski, R.: Stochastic vibrations of the Euler–Bernoulli beam based on various versions of the gradient nonlocal elasticity theory. Prob. Eng. Mech. **56**, 27–34 (2019)
15. Chen, H., Guirao, J., Cao, D.Q., Jiang, J., Fan, X.: Stochastic Euler–Bernoulli beam driven by additive white noise: global random attractors and global dynamics. Nonlinear Anal. **185**, 216–246 (2019)
16. Zuyev, A., Kucher, J.: Stabilization of a flexible beam model with distributed and lumped controls (in Russian). Dyn. Syst. **3**(31), 25–35 (2013). http://dynsys.cfuv.ru/wp-content/uploads/2016/12/002zuyev.pdf
17. Zuyev, A.L.: Partial Stabilization and Control of Distributed Parameter Systems with Elastic Elements. Springer, Cham (2015)
18. Baiocchi, C., Komornik, V., Loreti, P.: Ingham-Beurling type theorems with weakened gap conditions. Acta Math. Hungar. **97**, 55–95 (2002)
19. Krabs, W.: On Moment Theory and Controllability of One-Dimensional Vibrating Systems and Heating Processes. Springer, Berlin (1992)
20. Krabs, W., Sklyar, G.M.: On Controllability of Linear Vibrations. Nova Science Publishers, Hauppauge (2002)
21. Dullinger, C., Schirrer, A., Kozek, M.: Advanced control education: optimal & robust MIMO control of a flexible beam setup. IFAC Proc. Vol. **47**(3), 9019–9025 (2014)
22. Zuyev, A., Sawodny, O.: Stabilization and observability of a rotating Timoshenko beam model. Math. Problems Eng. **2007**, 1–19 (2007). https://doi.org/10.1155/2007/57238
23. Zuyev, A., Sawodny, O.: Stabilization of a flexible manipulator model with passive joints. IFAC Proc. Vol. **38**, 784–789 (2005)

Control Design for Linear Port-Hamiltonian Boundary Control Systems: An Overview

A. Macchelli, Y. Le Gorrec, H. Ramírez, H. Zwart, and F. Califano

Abstract In this paper, we provide an overview of some control synthesis methodologies for boundary control systems (BCS) in port-Hamiltonian form. At first, it is shown how to design a state-feedback control action able to shape the energy function to move its minimum at the desired equilibrium, and how to achieve asymptotic stability via damping injection. Secondly, general conditions that a linear regulator has to satisfy to have a well-posed and exponentially stable closed-loop system are presented. This second methodology is illustrated with reference to two specific stabilisation scenarios, namely when the plant is in impedance or in scattering form. It is also shown how these techniques can be employed in the analysis of more general systems described by coupled PDEs and ODEs. As an example, the repetitive control scheme is studied, and conditions to have asymptotic tracking of generic periodic reference signals are presented.

A. Macchelli (✉)
University of Bologna, Department of Electrical, Electronic and Information Engineering (DEI), Bologna, Italy
e-mail: alessandro.macchelli@unibo.it

Y. Le Gorrec
FEMTO-ST Institute, AS2M Department, University of Bourgogne-Franche-Comté/CNRS, Besançon, France
e-mail: legorrec@femto-st.fr

H. Ramírez
Universidad Técnica Federico Santa Maria, Departamento de Electrónica, Valparaíso, Chile
e-mail: hector.ramireze@usm.cl

H. Zwart
Department of Applied Mathematics, University of Twente, Enschede, The Netherlands
e-mail: h.j.zwart@utwente.nl

F. Califano
Robotics and Mechatronics Lab, University of Twente, Enschede, The Netherlands
e-mail: f.califano@utwente.nl

G. Sklyar, A. Zuyev (eds.), *Stabilization of Distributed Parameter Systems: Design Methods and Applications*, SEMA SIMAI Springer Series 2,
https://doi.org/10.1007/978-3-030-61742-4_4

1 Introduction

Port-Hamiltonian systems [16] have been introduced about 25 years ago to describe lumped parameter physical systems in an unified manner [6]. The generalisation to the infinite- dimensional scenario led to the definition of distributed port-Hamiltonian systems [26], introduced about 15 years ago. Most of the current research on stabilisation techniques deal with the development of boundary controllers, see e.g. [1, 11–15, 20, 22, 23]. In this paper, we illustrate the basic control design techniques for a particular class of linear, infinite-dimensional port-Hamiltonian systems with one-dimensional domain, and boundary actuation and sensing. As proved in [10], such systems are boundary control systems (BCS) in the sense of the semigroup theory [4], and have been studied in detail in [9].

The simplest way of designing such boundary controllers is to add some dissipation at the boundary (damping injection), and use the total energy as Lyapunov function to prove asymptotic/exponential stability of the zero equilibrium state. A more sophisticated approach consists in adding a further step (energy-shaping), in which the closed-loop energy function is shaped to shift its equilibrium; stability is assured by the passivity of the closed-loop system. Two possible implementations of such technique are presented here. In the first one, the energy-shaping task is accomplished by generating a set of invariants (Casimir functions) that relate the state of the BCS to the state of the dynamical controller, [13, 14, 19, 22–24], and the shape of the closed-loop energy function is changed by acting on the Hamiltonian of the controller itself. The main drawback is that it is not possible to deal with equilibria that require an infinite amount of supplied energy in steady state (dissipation obstacle). This limitation is solved by the second approach. The idea is to mimic the energy-Casimir method without requiring the existence of invariants, and going through dynamic extension/reduction. The control action is selected among all the possible state-feedback laws able to shape the closed-loop Hamiltonian function e.g. to have an isolated minimum at the equilibrium. In this way, simple stability is obtained and, to have asymptotic stability, it is necessary to add damping. The result is that the final system is asymptotically stable, [15].

A second general approach for control design consists in determining the conditions that a control system has to meet so that the related closed-loop system is asymptotically/exponentially stable. The technique presented in this paper is an extension of [27] or, more precisely, of [20]. Differently, here the BCS is no longer required to be passive and the stability result can be applied to all the possible parametrisation of the input–output mapping presented in [10]. The resulting BCS turns out to be dissipative, and the control design follows two main steps. In the first one, conditions on the controller structure are obtained so that the system of coupled PDEs and ODEs associated with the closed-loop dynamics is a well-posed BCS. Then, in a second step, dissipation is added to let the closed-loop energy (storage) function decrease exponentially. This fact implies the exponential stability of the equilibrium, [12]. The potentialities of the approach are illustrated in case the port-Hamiltonian BCS is in impedance or in scattering form. In both cases, sufficient conditions that the finite- dimensional controller has to satisfy to have an exponentially stable closed-loop system are provided. The proposed methodology can be also applied for the analysis of dynamical systems resulting from the interconnection of sub-systems modelled by means of PDEs and ODEs. To illustrate this feature, the stability analysis of the repetitive control [8] in the linear case is presented.

2 Distributed Port-Hamiltonian Systems

In this paper, we refer to the class of linear distributed port-Hamiltonian systems on real Hilbert spaces studied in [9, 10, 20, 28], i.e. to systems described by the PDE

$$\frac{\partial x}{\partial t}(t,z) = P_1 \frac{\partial}{\partial z}\big(\mathcal{L}(z)x(t,z)\big) + (P_0 - G_0)\mathcal{L}(z)x(t,z) \tag{1}$$

with $x \in X := L^2(a, b; \mathbb{R}^n)$, and $z \in [a, b]$. Moreover, it is assumed that $P_1 = P_1^{\mathrm{T}}$ is invertible, $P_0 = -P_0^{\mathrm{T}}$, $G_0 = G_0^{\mathrm{T}} \geq 0$, and $\mathcal{L}(\cdot)$ is a bounded and Lipschitz continuous matrix-valued function such that $\mathcal{L}(z) = \mathcal{L}^{\mathrm{T}}(z)$ and $\mathcal{L}(z) \geq \kappa I$, with $\kappa > 0$, for all $z \in [a, b]$. For the sake of clarity, $(\mathcal{L}x)(t, z) := \mathcal{L}(z)x(t, z)$. We say that the symmetric matrix M is positive definite, in short $M > 0$, if all its eigenvalues are positive, and positive semi-definite, in short $M \geq 0$, if its eigenvalues are non-negative. The state space X is endowed with the inner product $\langle x_1 \mid x_2\rangle_{\mathcal{L}} = \langle x_1 \mid \mathcal{L}x_2\rangle$ and norm $\|x_1\|^2_{\mathcal{L}} = \langle x_1 \mid x_1\rangle_{\mathcal{L}}$, where $\langle \cdot \mid \cdot\rangle$ denotes the natural L^2-inner product. The selection of this space for the state variable is motivated by the fact that $\|\cdot\|^2_{\mathcal{L}}$ is strongly linked to the energy function of (1). As a consequence, X is also called the space of energy variables, and $\mathcal{L}x$ denote the co-energy variables.

Let $H^1(a, b; \mathbb{R}^n)$ denote the Sobolev space of order one. The PDE (1) can be compactly written as $\dot{x} = \mathcal{J}x$, where

$$\mathcal{J}x := P_1 \frac{\partial}{\partial z}(\mathcal{L}x) + (P_0 - G_0)\mathcal{L}x \tag{2}$$

is a linear operator with domain $D(\mathcal{J}) = \{\mathcal{L}x \in H^1(a, b; \mathbb{R}^n)\}$. To have a port-Hamiltonian system, such PDE is completed by the set of boundary port variables $f_\partial, e_\partial \in \mathbb{R}^n$ that are a linear combination of the restriction of the co-energy variables $\mathcal{L}x \in H^1(a, b; \mathbb{R}^n)$ to the boundary and are defined by

$$\begin{pmatrix} f_\partial \\ e_\partial \end{pmatrix} = \underbrace{\frac{1}{\sqrt{2}}\begin{pmatrix} P_1 & -P_1 \\ I & I \end{pmatrix}}_{=:R} \begin{pmatrix} (\mathcal{L}x)(b) \\ (\mathcal{L}x)(a) \end{pmatrix} \tag{3}$$

Theorem 1 *Let W be a full rank $n \times 2n$ real matrix, and define the input mapping $\mathcal{B}: H^1(a,b;\mathbb{R}^n) \to \mathbb{R}^n$ and the input $u(t)$ as*

$$u(t) = W \begin{pmatrix} f_\partial(t) \\ e_\partial(t) \end{pmatrix} =: \mathcal{B}x(t) \tag{4}$$

The operator $\bar{\mathcal{J}}x := P_1 \frac{\partial}{\partial z}(\mathcal{L}x) + (P_0 - G_0)\mathcal{L}x$ with domain

$$D(\bar{\mathcal{J}}) = \left\{ \mathcal{L}x \in H^1(a,b;\mathbb{R}^n) \mid \begin{pmatrix} f_\partial \\ e_\partial \end{pmatrix} \in \operatorname{Ker} W \right\} \tag{5}$$

generates a contraction semigroup on X if and only if

$$W \Sigma W^{\mathrm{T}} \geq 0, \qquad \Sigma = \begin{pmatrix} 0 & I \\ I & 0 \end{pmatrix} \tag{6}$$

and the system (1) with input (4) is a boundary control system on X, [4, Theorem 3.3.3], provided that $u \in C^2(0,\infty;\mathbb{R}^n)$. Note that (5) is equivalent to require that $u = \mathcal{B}x = 0$. Moreover, let $\tilde{W}$ be a full rank $n \times 2n$ matrix such that $\begin{pmatrix} W^{\mathrm{T}} & \tilde{W}^{\mathrm{T}} \end{pmatrix}$ is invertible, and let P_W be given by

$$P_W = \begin{pmatrix} W\Sigma W^{\mathrm{T}} & W\Sigma \tilde{W}^{\mathrm{T}} \\ \tilde{W}\Sigma W^{\mathrm{T}} & \tilde{W}\Sigma\tilde{W}^{\mathrm{T}} \end{pmatrix}^{-1} = \begin{pmatrix} W \\ \tilde{W} \end{pmatrix}^{-\mathrm{T}} \Sigma \begin{pmatrix} W \\ \tilde{W} \end{pmatrix}^{-1} \tag{7}$$

Define the output as

$$y(t) = \tilde{W} \begin{pmatrix} f_\partial(t) \\ e_\partial(t) \end{pmatrix} =: \mathcal{C}x(t) \tag{8}$$

with $\mathcal{C}: H^1(a,b;\mathbb{R}^n) \to \mathbb{R}^n$. Then, for $(\mathcal{L}x)(0) \in H^1(a,b;\mathbb{R}^n)$, the following energy-balance equation is satisfied:

$$\frac{1}{2}\frac{\mathrm{d}}{\mathrm{d}t}\|x(t)\|^2_{\mathcal{L}} \leq \frac{1}{2}\begin{pmatrix} u(t) \\ y(t) \end{pmatrix}^{\mathrm{T}} P_W \begin{pmatrix} u(t) \\ y(t) \end{pmatrix} \tag{9}$$

Proof See [10, Theorem 4.1]. □

The energy-balance relation (9) shows that (1) with input–output mapping defined by (4) and (8) is a dissipative system, [2], with storage function $H(x) := \frac{1}{2}\|x\|^2_{\mathcal{L}}$, and supply rate

$$s(u,y) = \frac{1}{2}\begin{pmatrix} u \\ y \end{pmatrix}^{\mathrm{T}} P_W \begin{pmatrix} u \\ y \end{pmatrix} =: \frac{1}{2}\begin{pmatrix} u \\ y \end{pmatrix}^{\mathrm{T}} \begin{pmatrix} U & S \\ S^{\mathrm{T}} & Y \end{pmatrix} \begin{pmatrix} u \\ y \end{pmatrix} \tag{10}$$

where $U = U^{\mathrm{T}}$, and $Y = Y^{\mathrm{T}}$.

Remark 1 The input–output mapping of system (1) is in impedance form if W and $\tilde{W}$ in (4) and (8), respectively, are chosen such that $W\Sigma W^{\mathrm{T}} = \tilde{W}\Sigma\tilde{W}^{\mathrm{T}} = 0$ and $\tilde{W}\Sigma W^{\mathrm{T}} = I$, which leads to a supply rate (10) equal to $s(u, y) = y^{\mathrm{T}}u$. Differently, (1) is in scattering form if W and $\tilde{W}$ are such that $W\Sigma W^{\mathrm{T}} = -\tilde{W}\Sigma\tilde{W}^{\mathrm{T}} = I$ and $\tilde{W}\Sigma W^{\mathrm{T}} = 0$, which leads to a supply rate (10) equal to $s(u, y) = \frac{1}{2}u^{\mathrm{T}}u - \frac{1}{2}y^{\mathrm{T}}y$.

Before presenting the design methodologies, a preliminary problem is to understand when the linear system resulting from the feedback interconnection of (1) with a linear control system is well-posed, i.e. it is BCS. In this respect, let us consider the following linear control system:

$$\begin{cases} \dot{x}_c(t) = A_c x_c(t) + B_c u_c(t) \\ y_c(t) = C_c x_c(t) + D_c u_c(t) \end{cases} \tag{11}$$

where $x_c \in \mathbb{R}^{n_c}$ and $u_c, y_c \in \mathbb{R}^n$. It is assumed that A_c has eigenvalues with non-positive real part, and that the pair (A_c, B_c) is controllable. System (11) is interconnected to the boundary of (1) in standard feedback interconnection through the input $u(t)$ and $y(t)$ defined in (4) and (8), respectively, as shown in Fig. 1. This means that

$$\begin{pmatrix} u_c(t) \\ u(t) \end{pmatrix} = \begin{pmatrix} 0 & -I \\ I & 0 \end{pmatrix} \begin{pmatrix} y_c(t) \\ y(t) \end{pmatrix} + \begin{pmatrix} u_c'(t) \\ u'(t) \end{pmatrix} \tag{12}$$

where $u_c', u' \in \mathbb{R}^n$ are auxiliary signals. Finally, it is assumed that there exists a symmetric, positive definite $n_c \times n_c$ real matrix Q_c such that (11) is dissipative with storage function $H_c(x_c) := \frac{1}{2}x_c^{\mathrm{T}} Q_c x_c$ and supply rate

$$s_c(u_c, y_c) = \frac{1}{2}\begin{pmatrix} u_c \\ y_c \end{pmatrix}^{\mathrm{T}} \begin{pmatrix} U_c & S_c \\ S_c^{\mathrm{T}} & Y_c \end{pmatrix} \begin{pmatrix} u_c \\ y_c \end{pmatrix} \tag{13}$$

with $U_c = U_c^{\mathrm{T}}$, and $Y_c = Y_c^{\mathrm{T}}$.

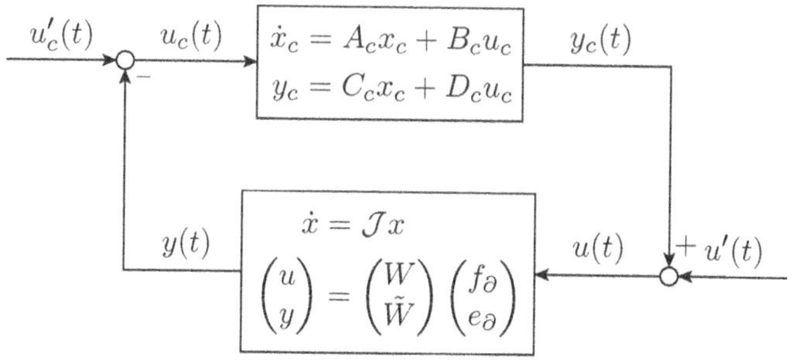

Fig. 1 Distributed port-Hamiltonian system (1) with boundary controller (11)

The closed-loop system resulting from the interconnection of (1) and (11) through the set of relations (12) can be compactly written as

$$\begin{cases} \dot{\xi}(t) = \mathcal{J}_{cl}\xi(t) + B_{cl}u_c'(t) \\ D_c u_c'(t) + u'(t) = \big(\mathcal{B} + D_c C \; -C_c\big)\,\xi(t) =: \mathcal{B}'\xi(t) \end{cases} \tag{14}$$

where the operators $\mathcal{B}$ and C are defined in (4) and (8), respectively, $\xi = (x, x_c) \in X_{cl} := X \times \mathbb{R}^{n_c}$ is the state variable, $\mathcal{J}_{cl} : D(\mathcal{J}_{cl}) \subset X_{cl} \to X_{cl}$ and $B_{cl} : \mathbb{R}^n \to X_{cl}$ are the linear operators

$$\mathcal{J}_{cl}\xi := \begin{pmatrix} \mathcal{J} & 0 \\ -B_c C & A_c \end{pmatrix} \begin{pmatrix} x \\ x_c \end{pmatrix} \qquad B_{cl}v := \begin{pmatrix} 0 \\ B_c v \end{pmatrix} \tag{15}$$

with

$$D(\mathcal{J}_{cl}) := D(\mathcal{J}) \times \mathbb{R}^{n_c} \tag{16}$$

being $\mathcal{J}$ the operator introduced in (2). Moreover, the state space X_{cl} is endowed with the inner product $\langle \xi_1 \mid \xi_2 \rangle_{X_{cl}} = \langle x_1 \mid x_2 \rangle_{\mathcal{L}} + x_{c,1}^{\mathrm{T}} Q_c x_{c,2}$. Some fundamental properties associated to the coupled PDEs and ODEs that describe the closed-loop dynamics are discussed in the next proposition.

Proposition 1 *Let us consider the closed-loop system resulting from the feedback interconnection (12) of (1) and (11), which results in (14). If*

$$\begin{pmatrix} Y & -S^{\mathrm{T}} \\ -S & U \end{pmatrix} + \begin{pmatrix} U_c & S_c \\ S_c^{\mathrm{T}} & Y_c \end{pmatrix} \leq 0 \tag{17}$$

the operator $\bar{\mathcal{J}}_{cl}$ defined as $\bar{\mathcal{J}}_{cl}\xi := \begin{pmatrix} \mathcal{J} & 0 \\ -B_c C & A_c \end{pmatrix} \begin{pmatrix} x \\ x_c \end{pmatrix}$ *with domain*

$$D(\bar{\mathcal{J}}_{cl}) = \left\{ \begin{pmatrix} x \\ x_c \end{pmatrix} \in X_{cl} \mid x \in D(\mathcal{J}),\ \text{and}\ \mathcal{B}' \begin{pmatrix} x \\ x_c \end{pmatrix} = 0 \right\} \tag{18}$$

and $\mathcal{B}'$ defined in (14) generates a contraction semigroup on X_{cl}. Moreover, (14) with $\mathcal{J}_{cl}$ and B_{cl} defined by (15) and (16) is a BCS on X_{cl} if $u_c', u' \in C^2(0, \infty; \mathbb{R}^n)$.

Proof This result is an extension of [12, Proposition 7]. □

Remark 2 If system (1) is in impedance form, see Remark 1, the control system (11) meets the condition of the previous proposition, for example, if it is passive, i.e. dissipative with respect to the supply rate $s_c(u_c, y_c) = y_c^{\mathrm{T}} u_c$. This result has been proved in [11, 27]. On the other hand, if (1) is in scattering form, the control system (11) can be selected such that it is dissipative with respect to the supply rate

$s_c(u_c, y_c) = \frac{1}{2}\gamma^2 \|u_c\|^2 - \frac{1}{2}\|y_c\|^2$, with $|\gamma| \le 1$. In other words, (11) should have a L^2-gain lower than γ, with $|\gamma| \le 1$, [25].

3 Energy-Shaping Design by Interconnection and State-Feedback

Proposition 1 shows when the system resulting from the feedback interconnection (12) of (1) and (11) is well-posed. It is also easy to check that such system is dissipative with storage function $H_{cl}(x, x_c) := H(x) + H_c(x_c)$, and the idea is to use H_{cl} as Lyapunov function. The control design procedure starts by guaranteeing that H_{cl} has a minimum at the desired equilibrium with a proper choice of H_c. As in the finite-dimensional case [18], if it is possible to find invariants of the form $C(x, x_c) = x_c - F(x)$ that do not depend on the Hamiltonian of the system, then on every invariant manifold $x_c - F(x) = \kappa$, with $\kappa \in \mathbb{R}$ a constant which depends on the initial condition, the closed-loop Hamiltonian may be written as $H_{cl}(x) = H(x) + H_c(F(x) + \kappa)$. Hence, the equilibrium now depends on the choice of H_c and, on the invariant manifold, H_{cl} is function of the state of (1) only. These invariants are called Casimir functions, and their definition is reported below, [5, 14]. For simplicity and with Remarks 1 and 2 in mind, we assume that (1) and (11) are passive, and that the latter one has a port-Hamiltonian structure, i.e.:

$$A_c = (J_c - R_c)Q_c \quad B_c = G_c - P_c \quad C_c = (G_c + P_c)^{\mathrm{T}} Q_c \quad D_c = M_c + S_c \tag{19}$$

where $J_c = -J_c^{\mathrm{T}}$, $M_c = -M_c^{\mathrm{T}}$, $R_c = R_c^{\mathrm{T}}$, and $S_c = S_c^{\mathrm{T}}$, and such that

$$\begin{pmatrix} R_c & P_c \\ P_c^{\mathrm{T}} & S_c \end{pmatrix} \ge 0 \tag{20}$$

Note that (11) with (19) and (20) is passive and, once interconnected to (1), leads to a well-posed closed-loop system in the sense of Proposition 1.

Definition 1 Consider the BCS of Proposition 1, and assume that $u' = u'_c = 0$ in (12) and that (11) is such that (19) and (20) hold. A function $C : X \times \mathbb{R}^{n_c} \to \mathbb{R}$ is a Casimir function if $\dot{C}(x(t), x_c(t)) = 0$ along the solutions for every possible choice of $\mathcal{L}(\cdot)$ and Q_c.

Proposition 2 *Under the conditions of Definition 1, the functional*

$$C(x(t), x_c(t)) := \Gamma^{\mathrm{T}} x_c(t) + \int_a^b \Psi^{\mathrm{T}}(z) x(t, z)\, \mathrm{d}z \tag{21}$$

with $\Gamma \in \mathbb{R}^{n_c}$ *and* $\Psi \in H^1(a,b;\mathbb{R}^n)$, *is a Casimir function for the closed-loop system if and only if*

$$
\begin{aligned}
P_1 \frac{\mathrm{d}\Psi}{\mathrm{d}z}(z) + (P_0 + G_0)\Psi(z) = 0 \\
(J_c + R_c)\Gamma + (G_c + P_c)\tilde{W} R \begin{pmatrix} \Psi(b) \\ \Psi(a) \end{pmatrix} = 0 \\
(G_c + P_c)^{\mathrm{T}}\Gamma + [W + (M_c - S_c)\tilde{W}] R \begin{pmatrix} \Psi(b) \\ \Psi(a) \end{pmatrix} = 0
\end{aligned}
\tag{22}
$$

From the first condition in (22) we can show that it is always possible to find n independent Casimir functions; so, in (11), we can assume that $n_c = n$. Now, let $\hat{\Gamma} := (\Gamma_1, \dots, \Gamma_n)$ and $\hat{\Psi} := (\Psi_1, \dots, \Psi_n)$ be the $n \times n$ matrices built from the elements that appear in each Casimir function (21). If $\hat{\Gamma}$ is invertible, under the conditions of Proposition 2, we have that

$$
x_c(t) = -\hat{\Gamma}^{-1} \int_a^b \hat{\Psi}^{\mathrm{T}}(z) x(t,z)\,\mathrm{d}z + \kappa \tag{23}
$$

with $\kappa \in \mathbb{R}^n$ a constant that depends only on the initial condition of (1) and (11). As a consequence, the controller Hamiltonian H_c is in fact a function of the state variables of the plant, and may be chosen to obtain a desired shape for the closed-loop energy. In the linear case, we have that $H_c(x_c) := \frac{1}{2} x_c^{\mathrm{T}} Q_c x_c$, with $Q_c = Q_c^{\mathrm{T}} > 0$. Note that it is also possible to project the state of the closed-loop system (x, x_c), on that state space of the plant to write the control action in state-feedback form, i.e.:

$$
\begin{aligned}
u(t) &= C_c x_c(t) - D_c y(t) + u'(t) \\
&= (G_c + P_c)^{\mathrm{T}} Q_c x_c(t) - (M_c + S_c) y(t) + u'(t)
\end{aligned}
\tag{24}
$$

where the output equation of (11) has been taken into account. In a state-feedback realisation of (24), the expression of $x_c(t)$ is given by (23), in which κ can be conveniently chosen equal to 0. This step is usually named reduction step because it reduces the dynamic contribution of the controller to a static one using a state-feedback approach. It is possible to verify that the final closed-loop dynamic is given by the following BCS:

$$
\begin{aligned}
\frac{\partial x}{\partial t}(t,z) &= P_1 \frac{\partial}{\partial z}\frac{\delta H_{cl}}{\delta x}(x(t,z)) + (P_0 - G_0)\frac{\delta H_{cl}}{\delta x}(x(t,z)) \\
u'(t) &= W' R \begin{pmatrix} \frac{\delta H_{cl}}{\delta x}(x(t,b)) \\ \frac{\delta H_{cl}}{\delta x}(x(t,a)) \end{pmatrix}
\end{aligned}
\tag{25}
$$

in which δ is the variational derivative [17], $H_{cl}(x) := \frac{1}{2}\|x\|^2_{\mathcal{L}} + \frac{1}{2}x_c^{\mathrm{T}}(x)Q_c x_c(x)$ and W' is a $n \times 2n$ full rank matrix which satisfies the hypothesis of Theorem 1, i.e. $W'\Sigma W'^{\mathrm{T}} \geq 0$. Independently from the way in which the control action is implemented (dynamic extension or state-feedback), the closed-loop system has the same structure of the plant (1), i.e. the same matrices P_1, P_0, and G_0, but a different (shaped) Hamiltonian, namely H_{cl}.

Even if this approach allows to shape the Hamiltonian function of (1) into H_{cl} and to obtain the closed-loop system (25), the presence of dissipation imposes strong constraints on the applicability of the method itself. In particular, (22) implies that

$$G_0\hat{\Psi}(z) = 0 \qquad \begin{pmatrix} R_c & P_c \\ P_c^{\mathrm{T}} & S_c \end{pmatrix} \begin{pmatrix} \hat{\Gamma} \\ \tilde{W}R\begin{pmatrix} \hat{\Psi}(b) \\ \hat{\Psi}(a) \end{pmatrix} \end{pmatrix} = 0 \tag{26}$$

Such conditions, and the first one in particular, show that it is not possible to shape the closed-loop Hamiltonian in the coordinates in which dissipation is present. This limitation is called “dissipation obstacle”, and is related to the fact that the (passive) control system (11) has just a finite amount of energy at disposal to drive the state of the plant (1) towards the desired equilibrium. To overcome this limitation, the idea is to start from the feedback law (23) and (24) derived in the context of the immersion/reduction scheme and to directly design a state feedback law that shapes the closed-loop Hamiltonian function, but without relying on the dynamic extension and on the Casimir functions. Asymptotic stability is then guaranteed by damping injection via the auxiliary input $u'(t)$.

Proposition 3 *Consider the system (1) with boundary input u defined in (4). The feedback law $u(t) = \beta(x(t)) + u'(t)$ in which u' is an auxiliary input, maps (1) into*

$$\begin{aligned} \frac{\partial x}{\partial t}(t,z) &= P_1 \frac{\partial}{\partial z}\frac{\delta H_d}{\delta x}(x(t,z)) + (P_0 - G_0)\frac{\delta H_d}{\delta x}(x(t,z)) \\ u'(t) &= WR\begin{pmatrix} \frac{\delta H_d}{\delta x}(x(t,b)) \\ \frac{\delta H_d}{\delta x}(x(t,a)) \end{pmatrix} \end{aligned} \tag{27}$$

with $H_d(x) := \frac{1}{2}\|x\|^2_{\mathcal{L}} + H_a(\xi(x))$ and H_a an arbitrary C^1 function, if

$$\beta(x) = -WR\begin{pmatrix} \Phi(b) \\ \Phi(a) \end{pmatrix} \frac{\partial H_a}{\partial \xi}(\xi(x)) \tag{28}$$

in which $\xi(x(t,\cdot)) := \int_a^b \Phi^{\mathrm{T}}(z)x(t,z)\,\mathrm{d}z$, being $\Phi(z) := (\Phi_1(z), \ldots, \Phi_n(z))$ and each $\Phi_i \in H^1(a,b;\mathbb{R}^n)$ independent solution of

$$P_1\frac{\mathrm{d}\Phi_i}{\mathrm{d}z}(z) + (P_0 - G_0)\Phi_i(z) = 0 \tag{29}$$

Proof This result is a reformulation of [15, Proposition 4.1 and Lemma 4.2]. □

From the previous proposition, it is not clear how to select $H_a(\xi)$ and then $\beta(x)$ so that the energy function of (1) is properly shaped, for example to move the minimum at the desired equilibrium configuration $\mathcal{L}x^\star \in H^1(a, b; \mathbb{R}^n)$. Note that, due to the definition of $\Phi(z)$ in Proposition 3, there exists a unique $\phi \in \mathbb{R}^n$ such that $(\mathcal{L}x^\star)(z) = \Phi(z)\phi$. A possible choice for $H_a(\xi)$ is then

$$H_a(\xi) := \frac{1}{2}(\xi - \xi^\star)^{\mathrm{T}} Q_a (\xi - \xi^\star) - \phi^{\mathrm{T}} \xi + H_a^\star \tag{30}$$

where $\xi^\star := \xi(x^\star)$, $H_a^\star$ is a constant selected so that $H_a(\xi) > 0$ for all $\xi \neq 0$, and $Q_a = Q_a^{\mathrm{T}} > 0$. With this choice, we have that

$$\frac{\delta H_d}{\delta x}(x(t,z)) = \mathcal{L}(z)x(t,z) + \Phi(z) Q_a \big[\xi(x(t,z)) - \xi_\star\big] - \Phi(z)\phi \tag{31}$$

that is equal to 0 when $x(t,z) = x^\star(z)$ because of the definition of ϕ. Such critical "point" is isolated, and is a minimum for H_d. From (30) and thanks to (28), the corresponding energy-shaping control law $\beta(x)$ can be obtained. It is worth noticing that the same procedure can be applied for the selection of H_c once a set of invariants has been computed by following Proposition 2. In fact, for any $\hat{\Psi}(z)$ in (23), from (22) and (26), it is possible to check that $P_1 \frac{\mathrm{d}\hat{\Psi}}{\mathrm{d}z}(z) + (P_0 - G_0)\hat{\Psi}(z) = 0$, which is the same condition that $\Phi(z)$ introduced in Proposition 3 has to meet. This means that H_c can be selected equal to H_a, and so $H_{cl} \equiv H_d$. In other words and similarly to the finite-dimensional case [18], the energy-shaping methodology based on Proposition 2 is a particular case of the procedure illustrated in Proposition 3. The difference is that in the latter case, it is possible to shape the Hamiltonian in the coordinates that have pervasive dissipation. Differently, not all the equilibria of (1) can be stabilised with the first approach.

In any case, once the closed-loop Hamiltonian has been defined, and these considerations are valid either for H_{cl} in (25) and H_d in (26), convergence of the trajectories towards the new minimum of the energy function is obtained by introducing dissipation via the auxiliary input $u'(t)$. With an eye on the energy-shaping procedure presented in Proposition 3 and on the closed-loop dynamic (26), by Theorem 1, a natural choice for the output dual to $u'(t)$ is

$$y'(t) = \tilde{W} R \begin{pmatrix} \frac{\delta H_d}{\delta x}(x(t,b)) \\ \frac{\delta H_d}{\delta x}(x(t,a)) \end{pmatrix} \tag{32}$$

for which it is immediate to check that $\frac{\mathrm{d}}{\mathrm{d}t} H_d(x(t)) \leq y'^{\mathrm{T}}(t) u'(t)$. A simple way to introduce dissipation is by imposing that, [18]:

$$u'(t) = -\Xi y'(t), \qquad \Xi = \Xi^{\mathrm{T}} \geq 0 \tag{33}$$

Proposition 4 *Consider the linear BCS of Theorem 1, and the equilibrium state* $x^\star(z)$. *Then, the control action* $u(t) = \beta(x(t,\cdot)) + u'(t)$ *in which* β *is defined as in (28) with the choice (30) for* H_a, *and* u' *as in (33) with* $\Xi > 0$, *makes* $x^\star(z)$ *asymptotically stable.*

Proof See [15, Theorem 4.5]. □

4 Exponential Stabilisation of Port-Hamiltonian Linear BCS

In the previous section, we have shown that via energy-shaping and damping injection it is possible to asymptotically stabilise an equilibrium configuration for the BCS of Theorem 1. With an eye on the feedback scheme reported in Fig. 1, the aim is now to show how it is possible to choose the linear control system (11) so that the closed-loop system is not only well-posed in the sense of Proposition 1, but also exponentially stable. The result generalises what has been presented in [20] for port-Hamiltonian BCS in impedance form, under the further requirement that (11) is a port-Hamiltonian system.

Let us assume that the linear control system (11) is such that Proposition 1 holds. The main requirement is that the following LMI holds true:

$$\begin{pmatrix} Q_c A_c + A_c^{\mathrm{T}} Q_c & Q_c B_c \\ B_c^{\mathrm{T}} Q_c & 0 \end{pmatrix} - \begin{pmatrix} C_c^{\mathrm{T}} Y_c C_c & C_c^{\mathrm{T}} Y_c D_c \\ D_c^{\mathrm{T}} Y_c C_c & U_c + D_c^{\mathrm{T}} Y_c D_c \end{pmatrix} - \\ - \begin{pmatrix} 0 & C_c^{\mathrm{T}} S_c \\ S_c^{\mathrm{T}} C_c & D_c^{\mathrm{T}} S_c + S_c^{\mathrm{T}} D_c \end{pmatrix} \leq - \begin{pmatrix} -\delta_x (Q_c A_c + A_c^{\mathrm{T}} Q_c) & 0 \\ 0 & \delta_u I \end{pmatrix} \tag{34}$$

with δ_x and δ_u two positive constants. When $\delta_x = \delta_u = 0$, (34) states that (11) is dissipative with storage function $H_c(x_c) := \frac{1}{2} x_c^{\mathrm{T}} Q_c x_c$, with $Q_c = Q_c^{\mathrm{T}} > 0$, and supply rate (13). From a physical point of view, δ_x is related to the presence of internal damping in the control system responsible for attenuating the lower frequencies in the plant dynamics. Differently, δ_u assures that the higher frequencies are damped.

Proposition 5 *Under the same conditions of Proposition 1, assume that the control system (11) is such that* A_c *has all the eigenvalues with negative real part, the pair* (A_c, B_c) *is controllable, and (33) holds with* $\delta_x > 0$ *and* $\delta_u > 0$. *Then, the closed-loop system (14) with* $u_c'(t) = u'(t) = 0$ *is exponentially stable.*

Proof See [12, Proposition 12]. □

The previous result is now used to study the stability of systems whose dynamics are given in terms of coupled PDEs and ODEs. With Remark 1 in mind, we start with a standard regulation problem in which the distributed port-Hamiltonian system (1) is in impedance or in scattering form. Then, we focus on a different and apparently unrelated topic, i.e. repetitive control of linear systems, [8]. The goal is to determine the class of linear systems for which this control technique can be applied. Now, let us first assume that (1) is in impedance form, which implies that (1) is passive, i.e. dissipative with storage function given by the total energy $\frac{1}{2}\|x\|^2_{\mathcal{L}}$, and supply rate $s(u, y) = y^{\mathrm{T}}u$, where input $u(t)$ and output $y(t)$ are given in (4) and (8), respectively. From (34) in Proposition 5, the control system (11) leads to an exponentially stable closed-loop system if there exists $Q_c = Q_c^{\mathrm{T}} > 0$, $\delta_x > 0$ and $\delta_u > 0$ such that

$$\begin{pmatrix} (1-\delta_x)(Q_cA_c + A_c^{\mathrm{T}}Q_c) & Q_cB_c \\ B_c^{\mathrm{T}}Q_c & 0 \end{pmatrix} - \begin{pmatrix} 0 & C_c^{\mathrm{T}} \\ C_c & D_c + D_c^{\mathrm{T}} - \delta_u I \end{pmatrix} \leq 0 \tag{35}$$

This implies that $D_c + D_c^{\mathrm{T}} \geq \delta_u I > 0$, and that $\dot{H}_c(x_c(t)) \leq y_c^{\mathrm{T}}(t)u_c(t) - \delta_u\|u_c(t)\|^2$, where $H_c(x_c) = \frac{1}{2}x_c^{\mathrm{T}}Q_cx_c$ is the storage function of (11). This relation implies that the control system has to be input strictly passive, [25].

Corollary 1 *Under the same conditions of Proposition 5, let us consider the port-Hamiltonian system (1), now assumed in impedance form. Moreover, let us denote by* $H_c(s) = C_c(sI - A_c)^{-1}B_c + D_c$ *the transfer matrix of (11). The closed-loop system (14) is exponentially stable if the linear system with transfer matrix* $H(s-\epsilon)$ *is strictly input passive for some* $\epsilon > 0$.

In [20], the same result of Corollary 1 has been proved in case the control system (20) is in a specific port-Hamiltonian form, i.e. if (19) holds with $P_c = 0$ and $M_c = 0$. It is easy to see that if $A_c = (J_c - R_c)Q_c$ is Hurwitz, $R_c \neq 0$, and $D_c + D_c^{\mathrm{T}} > 0$ then (35) holds, and the closed-loop system is exponentially stable. Note that (35) holds even for $\delta_x = 0$. In fact, from (19) we have that $C_c = G_c^{\mathrm{T}}Q_c = B_c^{\mathrm{T}}Q_c$, and $Q_cA_c + A_c^{\mathrm{T}}Q_c = -2Q_cR_cQ_c \leq 0$ because J_c is skew-symmetric and $R_c \neq 0$. It is worth mentioning that an extension to the case in which the control system is nonlinear has been presented in [21].

Analogous considerations can be drawn if (1) is in scattering form, i.e. when input and output are selected in such a way that the distributed parameter system is dissipative with storage function $\frac{1}{2}\|x\|^2_{\mathcal{L}}$ and supply rate $s(u, y) = \frac{1}{2}\|u\|^2 - \frac{1}{2}\|y\|^2$. From (35) in Proposition 5, the regulator (11) exponentially stabilises the BCS (1) if there exists $Q_c = Q_c^{\mathrm{T}} > 0$, and $\delta_x > 0$ and $\delta_u > 0$ such that

$$\begin{pmatrix} (1-\delta_x)(Q_cA_c + A_c^{\mathrm{T}}Q_c) & Q_cB_c \\ B_c^{\mathrm{T}}Q_c & 0 \end{pmatrix} + \begin{pmatrix} C_c^{\mathrm{T}}C_c & C_c^{\mathrm{T}}D_c \\ D_c^{\mathrm{T}}C_c & D_c^{\mathrm{T}}D_c - (\gamma^2 - \delta_u)I \end{pmatrix} \leq 0 \tag{36}$$

for some γ such that $|\gamma| \leq 1$. The LMI (36) implies that $D_c^{\mathrm{T}}D_c - (\gamma^2 - \delta_u)I \leq 0$, i.e. that $D_c^{\mathrm{T}}D_c - I < 0$, which means that the feedthrough gain has to be lower than 1 or, equivalently, that the dissipation inequality $\dot{H}_c(x_c(t)) \leq \frac{1}{2}\gamma^2\|u_c(t)\|^2 - \frac{1}{2}\|y_c(t)\|^2$ holds true with $|\gamma| < 1$.

Corollary 2 *Under the same conditions of Proposition 5, let us consider the port-Hamiltonian system (1), now assumed in scattering form. Moreover, let us denote by* $H_c(s) = C_c(sI - A_c)^{-1}B_c + D_c$ *the transfer matrix of (11). The closed-loop system (14) is exponentially stable if, for some* $\epsilon > 0$, *the linear system with transfer matrix* $H(s - \epsilon)$ *has* L^2*-gain* $\gamma < 1$.

The final contribution is now to show how the previous methodological results can be applied on different control problems, provided that the closed-loop system is described by a set of coupled PDEs and ODEs. The focus is on repetitive control, [8], a simple technique to let a dynamical system to track and/or reject periodic exogenous signals with a known time period T. Its effectiveness relies on the Internal Model Principle [7], and the main properties depend on a particular element reported in Fig. 2 and denoted by $C(s)$. Such dynamical system, called *repetitive compensator*, is a pure time delay T surrounded by a positive feedback loop that represents, from an Internal Model Principle point of view, a generator of any periodic signal whose period equals the amount of time in the delay.

The repetitive compensator can be described by means of a delay PDE. When $u(t) = 0$, the particular structure of the compensator causes the initial condition associated to the delay equation, i.e. an arbitrary function defined on $[0, T]$, to be periodically transported along the domain to generate the periodic signal $y(t)$. The pair (u, y) defines the input–output mapping of the system, with u and y that depend on the boundary conditions of the PDE. The repetitive compensator admits an interpretation in terms of a BCS in port-Hamiltonian form in the sense of Theorem 1 if we select for (1) $P_1 = -I$, $P_0 = G_0 = 0$, and $\mathcal{L}(z) = I$, with $z \in [0, T]$, and if $W = \sqrt{2}\begin{pmatrix} I & 0 \end{pmatrix}$ and $\tilde{W} = \frac{\sqrt{2}}{2}\begin{pmatrix} -I & I \end{pmatrix}$. Moreover, it is easy to check that it obeys to the following energy-balance relation:

$$\frac{1}{2}\frac{\mathrm{d}}{\mathrm{d}t}\|x(t)\|_2^2 = y^{\mathrm{T}}(t)u(t) + \frac{1}{2}u^{\mathrm{T}}(t)u(t) = \frac{1}{2}\begin{pmatrix} u(t) \\ y(t) \end{pmatrix}^{\mathrm{T}}\begin{pmatrix} I & I \\ I & 0 \end{pmatrix}\begin{pmatrix} u(t) \\ y(t) \end{pmatrix}$$

An immediate consequence is that it is possible to treat repetitive control within the port-Hamiltonian framework or, more precisely, to rely on the stability tools discussed in this section to determine under which conditions on the plant dynamics the closed-loop system depicted in Fig. 2 is exponentially stable.

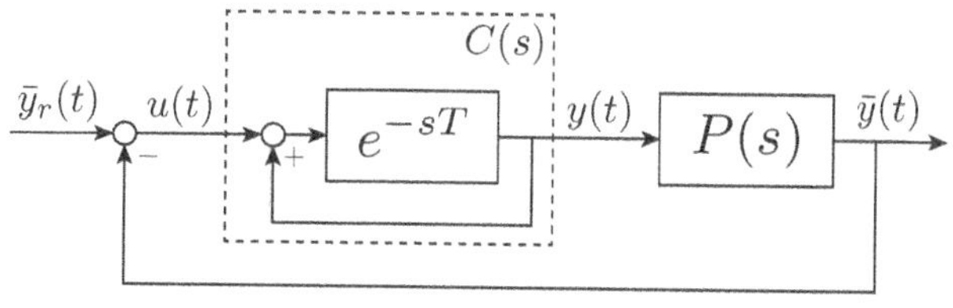

Fig. 2 Basic structure of (continuous-time) repetitive control, [8]

Since the repetitive compensator is a BCS, the scheme of Fig. 2 can be equivalently represented as in Fig. 1 with $\mathcal{J} = -\frac{\partial}{\partial z}$, $\bar{y}_r(t) \equiv u'(t)$ the periodic reference signal, $u_c'(t) = 0$, and under the hypothesis that the plant is the linear system (11). Note that, in this case, the distributed parameter system is the controller (repetitive compensator) responsible for stabilising on a periodic trajectory the finite- dimensional plant. From Proposition 1, existence of solution in closed-loop is guaranteed if (11) is dissipative with respect to the quadratic storage function $H_c(x_c) = \frac{1}{2}x_c^{\mathrm{T}} Q_c x_c$, and to the supply rate (13), in which $U_c = 0$, $S_c = I$, and $Y_c = -\sigma I$, with $\sigma \geq 1$. Then, Proposition 5 is instrumental to characterise the class of linear systems (11) for which the closed-loop system is exponentially stable. Then, the Internal Model Principle assures that the tracking error goes to zero in case of periodic reference signals $\bar{y}_r(t)$, [8].

Proposition 6 *The repetitive control scheme of Fig. 2 is well-posed and exponentially stable if the plant $P(s)$ takes the from (11), and it is such that A_c is Hurwitz, the pair (A_c, B_c) is controllable, and*

$$\begin{pmatrix} Q_c A_c + A_c^{\mathrm{T}} Q_c & Q_c B_c \\ B_c^{\mathrm{T}} Q_c & 0 \end{pmatrix} - \begin{pmatrix} -\sigma C_c^{\mathrm{T}} C_c & C_c^{\mathrm{T}} (I - \sigma D_c) \\ (I - \sigma D_c^{\mathrm{T}}) C_c & D_c^{\mathrm{T}} + D_c - \sigma D_c^{\mathrm{T}} D_c \end{pmatrix} \leq$$

$$\leq - \begin{pmatrix} -\delta_x \left(Q_c A_c + A_c^{\mathrm{T}} Q_c\right) & 0 \\ 0 & \delta_u I \end{pmatrix} \tag{37}$$

holds for a $Q_c = Q_c^{\mathrm{T}} > 0$ $\sigma \geq 1$, $\delta_x > 0$ and $\delta_u > 0$.

Proof The result follows from Proposition 5, in which the supply rate of (11) is given as in (13), with $U_c = 0$, $S_c = I$, and $Y_c = -\sigma I$. □

The property summarised in the previous proposition is consistent with the classical stability conditions of repetitive control. In fact, a necessary condition for (37) to hold true is that $D_c^{\mathrm{T}} + D_c - \sigma D_c^{\mathrm{T}} D_c \geq \delta_u I$, for all $\sigma \geq 1$, and $\delta_u > 0$. If for simplicity $D_c = \gamma I$, we have that $\sigma\gamma^2 - 2\gamma < 0$, i.e. that $0 < \gamma < \frac{2}{\sigma}$. So, it is necessary that (11) is strictly proper, and that the feedthrough gain γ is positive and lower than 2, which corresponds to $\sigma = 1$. Note that, since now the stability condition is given in time-domain and based on energy-considerations, it can be extended also to deal with nonlinear systems. A first attempt in this direction has been illustrated in [3]. Moreover, from Proposition 1, since $\sigma \geq 1$, it can be deduced that, to obtain a closed-loop system whose evolution is described in terms of a contraction C_0-semigroup, (11) has to be ν-output strictly passive [25], with $\nu \geq \frac{1}{2}$. Then, (37) forces (11) to have a non-null feedthrough term, and low frequency dissipation to guarantee exponential stability. With Corollaries 1 and 2 in mind, this latter condition is equivalent to require that the linear system with transfer matrix $H(s - \epsilon)$ is ν-output strictly passive with $\nu \geq \frac{1}{2}$, where $H(s)$ is the transfer matrix of (11).

5 Conclusions and Future Works

The goal of the paper is to present in an unified manner some of the basic control design techniques developed so far for linear BCS in port-Hamiltonian form characterised by a 1D spatial domain. The first synthesis methodology is based on state-feedback and capable to shape the energy function to move its minimum at the desired equilibrium. In this case, asymptotic stability is obtained via damping injection. The second technique, instead, provides some general conditions in terms of an LMI that a linear regulator has to satisfy to obtain a closed-loop system that is well-posed and exponentially stable. This methodology is quite general and powerful, and it has been illustrated with reference to two stabilisation scenarios, i.e. when the plant is in impedance or in scattering form. Moreover, because of its generality, it can be employed in the analysis of systems described by coupled PDEs and ODEs. As an example, the repetitive control scheme is studied.

Future researches are mainly focused to the extension of such results to BCS in port-Hamiltonian form in which the spatial domain is 2D or 3D. Another stimulating research topic deals with nonlinear BCS. In this respect, some preliminary results have been discussed in [3, 21], but some efforts are still required to develop a general theory that is applicable to a large class of systems.

Acknowledgments Héctor Ramírez gratefully acknowledges the support of the FONDECYT project with reference code 1191544 and of the BASAL project with reference code FB0008.

References

1. Augner, B., Jacob, B.: Stability and stabilization of infinite-dimensional linear port-Hamiltonian systems. Evol. Equ. Control Theory **3**(2), 207–229 (2014)
2. Brogliato, B., Lozano, R., Maschke, B., Egeland, O.: Dissipative Systems Analysis and Control. Communications and Control Engineering, 2nd edn. Springer, London (2007)
3. Califano, F., Bin, M., Macchelli, A., Melchiorri, C.: Stability analysis of nonlinear repetitive control schemes. IEEE Control Syst. Lett. **2**(4), 773–778 (2018)
4. Curtain, R., Zwart, H.: An Introduction to Infinite Dimensional Linear Systems Theory. Springer, New York (1995)
5. Dalsmo, M., van der Schaft, A.: On representation and integrability of mathematical structures in energy-conserving physical systems. SIAM J. Control Optim. **37**, 54–91 (1999)
6. Duindam, V., Macchelli, A., Stramigioli, S., Bruyninckx, H.: Modeling and Control of Complex Physical Systems: The Port-Hamiltonian Approach. Springer, Berlin (2009)
7. Francis, B., Wonham, W.: The internal model principle of control theory. Automatica **12**(5), 457–465 (1976)
8. Hara, S., Yamamoto, Y., Omata, T., Nakano, M.: Repetitive control system: a new type servo system for periodic exogenous signals. IEEE Trans. Autom. Control **33**(7), 659–668 (1988)
9. Jacob, B., Zwart, H.: Linear port-Hamiltonian systems on infinite-dimensional spaces. In: Operator Theory: Advances and Applications, vol. 223. Birkhäuser, Basel (2012)
10. Le Gorrec, Y., Zwart, H., Maschke, B.: Dirac structures and boundary control systems associated with skew-symmetric differential operators. SIAM J. Control Optim. **44**(5), 1864–1892 (2005)

11. Macchelli, A.: Boundary energy shaping of linear distributed port-Hamiltonian systems. Eur. J. Control **19**(6), 521–528 (2013)
12. Macchelli, A., Califano, F.: Dissipativity-based boundary control of linear distributed port-Hamiltonian systems. Automatica **95**, 54–62 (2018)
13. Macchelli, A., Melchiorri, C.: Control by interconnection of mixed port Hamiltonian systems. IEEE Trans. Autom. Control **50**(11), 1839–1844 (2005)
14. Macchelli, A., Melchiorri, C.: Modeling and control of the Timoshenko beam. The distributed port Hamiltonian approach. SIAM J. Control Optim. **43**(2), 743–767 (2005)
15. Macchelli, A., Le Gorrec, Y., Ramírez, H., Zwart, H.: On the synthesis of boundary control laws for distributed port-Hamiltonian systems. IEEE Trans. Autom. Control **62**(4), 1700–1713 (2017)
16. Maschke, B., van der Schaft, A.: Port controlled Hamiltonian systems: modeling origins and system theoretic properties. In: Proceedings of the 3rd IFAC Symposium on, Nonlinear Control Systems (NOLCOS 1992), pp. 282–288. Bordeaux (1992)
17. Olver, P.: Application of Lie Groups to Differential Equations, 2nd edn. Springer, Berlin (1993)
18. Ortega, R., van der Schaft, A., Mareels, I., Maschke, B.: Putting energy back in control. In: Control Systems Magazine, pp. 18–33. IEEE, Piscataway (2001)
19. Pasumarthy, R., van der Schaft, J.: Achievable Casimirs and its implications on control by interconnection of port-Hamiltonian systems. Int. J. Control **80**(9), 1421–1438 (2007)
20. Ramírez, H., Le Gorrec, Y., Macchelli, A., Zwart, H.: Exponential stabilization of boundary controlled port-Hamiltonian systems with dynamic feedback. IEEE Trans. Autom. Control **59**(10), 2849–2855 (2014)
21. Ramírez, H., Zwart, H., Le Gorrec, Y.: Stabilization of infinite dimensional port-Hamiltonian systems by nonlinear dynamic boundary control. Automatica **85**, 61–69 (2017)
22. Rodriguez, H., van der Schaft, A., Ortega, R.: On stabilization of nonlinear distributed parameter port-controlled Hamiltonian systems via energy shaping. In: Proceedings of the 40th IEEE Conference on Decision and Control (CDC 2001), vol. 1, pp. 131–136 (2001)
23. Schöberl, M., Siuka, A.: On Casimir functionals for infinite-dimensional port-Hamiltonian control systems. IEEE Trans. Autom. Control **58**(7), 1823–1828 (2013)
24. Siuka, A., Schöberl, M., Schlacher, K.: Port-Hamiltonian modelling and energy-based control of the Timoshenko beam. Acta Mech. **222**(1–2), 69–89 (2011)
25. van der Schaft, A.: L_2-Gain and Passivity Techniques in Nonlinear Control. Communication and Control Engineering, 3rd edn. Springer, London (2017)
26. van der Schaft, A., Maschke, B.: Hamiltonian formulation of distributed parameter systems with boundary energy flow. J. Geom. Phys. **42**(1–2), 166–194 (2002)
27. Villegas, J., Zwart, H., Le Gorrec, Y., Maschke, B., van der Schaft, A.: Stability and stabilization of a class of boundary control systems. In: Proceedings of the 44th IEEE Conference on Decision and Control and European Control Conference (CDC-ECC 2005), pp. 3850–3855 (2005)
28. Villegas, J., Zwart, H., Le Gorrec, Y., Maschke, B.: Exponential stability of a class of boundary control systems. IEEE Trans. Autom. Control **54**(1), 142–147 (2009)

Nonlinear Control of Continuous Fluidized Bed Spray Agglomeration Processes

Eric Otto, Stefan Palis, and Achim Kienle

Abstract Fluidized bed spray agglomeration is a complex particle formation process widely used in the agricultural, food, and pharmaceutical industry. It can described mathematically by population balance equations. This chapter deals with controlling the nonlinear partial integro-differential equation. Therefore, discrepancy based control, which guarantees exponential stability with respect to some generalized distance measure, is introduced. Conditions for convergence in a norm are discussed. Furthermore, robustness with respect to model uncertainties is shown.

1 Introduction

Fluidized bed spray agglomeration (FBSA) is an industrial particle formation process with the goal of producing particles with predefined properties. Hereby, two or more so-called primary particles are combined to form a new particle with different properties. For this purpose, a particle bed is fluidized within an upward air stream while a binder solution is sprayed into the process chamber wetting the particles. Due to particle collision and drying of the liquid layer, solid bridges between particles are formed. This process is depicted schematically in Fig. 1.

Important examples of products in particulate form are fertilizers in the agricultural industry, milk powder in the food industry, or medicals in the pharmaceutical

E. Otto
Otto von Guericke University Magdeburg, Magdeburg, Germany
e-mail: eric.otto@ovgu.de

S. Palis (✉)
Otto von Guericke University Magdeburg, Magdeburg, Germany

National Research University "Moscow Power Engineering Institute", Moscow, Russia
e-mail: stefan.palis@ovgu.de

A. Kienle
Otto von Guericke University Magdeburg, Max Planck Institute for Dynamics of Complex Technical Systems Magdeburg, Magdeburg, Germany

G. Sklyar, A. Zuyev (eds.), *Stabilization of Distributed Parameter Systems: Design Methods and Applications*, SEMA SIMAI Springer Series 2,
https://doi.org/10.1007/978-3-030-61742-4_5

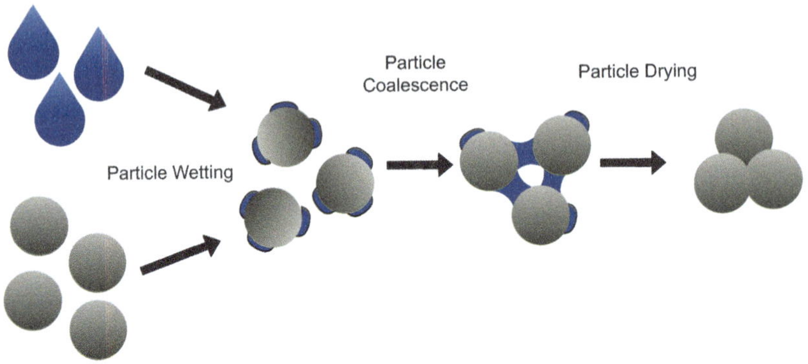

Fig. 1 Three stages of agglomeration: (a) particle wetting, (b) particle coalescence, and (c) particle drying

industry [1]. Particle properties such as size and shape, porosity, and flowability determine the quality of the product and its suitability in the subsequent processing. Model-based process control is one way to achieve a constant production rate on the one hand and the desired product quality on the other hand. Two factors are currently limiting the effective use of process control. Firstly appropriate process models capturing the important dynamics are hardly available and secondly deriving stabilizing controllers is a major challenge due to the complexity of the nonlinear infinite dimensional system description. In this contribution discrepancy based control [9, 10, 12] is used to design a stable closed-loop system for a simple process model. Here, stability with respect to a generalized distance measure is considered. Furthermore, conditions for stability with respect to a norm are discussed.

2 Process Modeling

The technical realization of the FBSA process is depicted schematically in Fig. 2. In continuous operation mode, new primary particles as well as binder solution are fed constantly while particles exceeding a predefined size are withdrawn. Advantages of this mode of operation are higher throughputs compared to batch agglomeration.

The standard approach for mathematical modeling of FBSA is using a population balance equation (PBE) balancing the number density distribution (NDD) $n(t, \mathbf{x})$ depending on time $t \in \mathscr{R}_{\geq 0}$ and some internal or external coordinates $\mathbf{x} \in \mathscr{R}^n$. The external coordinates are usually the spatial coordinates describing the position in three-dimensional space. However, under the assumption of ideal mixing of the particle population in the fluidized bed, the spatial distribution of particles can be lumped. Typical candidates for internal coordinates are particle properties such as particle size, porosity, or shape. A standard simplifying assumption for the modeling of agglomeration processes [5] is that particles are spherical, therefore

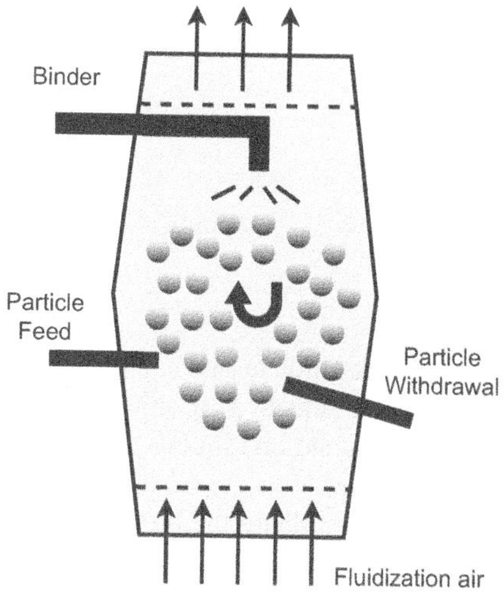

Fig. 2 Process scheme

their shape can be neglected. Additionally, further particle properties except for the characteristic volume $v \in \mathscr{R}_{\geq 0}$ are neglected, i.e. $\mathbf{x} = v$. The population balance equation is thus given as

$$\frac{\partial n(t, v)}{\partial t} = \dot{n}_{\mathrm{a}}(t, v) + \dot{n}_{\mathrm{f}}(t, v) - \dot{n}_{\mathrm{o}}(t, v) \tag{1}$$

with the aggregation term $\dot{n}_{\mathrm{a}}$ describing the formation of new particles, the feed term $\dot{n}_{\mathrm{f}}$ describing the particles added to the process, and the output term $\dot{n}_{\mathrm{o}}$ describing the withdrawn particles. In the following the terms are described in detail.

The aggregation term was originally derived by Hulburt and Katz [5] and consists of a birth and a death term

$$\dot{n}_{\mathrm{a}}(t, v) = B(t, v) - D(t, v). \tag{2}$$

The agglomeration rate

$$r = \beta(t, u, v - u)n(t, u)n(t, v - u) \tag{3}$$

with the agglomeration kernel $\beta(t, u, v)$ describes the number of agglomeration events per unit of time. Usually $\beta(t, u, v)$ is divided into a size-independent part $\beta_0(t)$ called the agglomeration efficiency and a size-dependent part $\beta(u, v)$ called the coalescence kernel. An agglomeration event is defined as collision and coalescence of two particles with volume u and $v - u$, forming a new particle of volume v. The agglomeration kernel can be interpreted as the frequency of particles aggregating per unit of time depending on the coordinate v. In the literature a variety of both empirical and analytical coalescence kernels have been proposed. A selection can be found in Table 1. As has been investigated in Bück et. al. [2],

Table 1 Selected coalescence kernels

Name	$\beta(u, v)$
Size-independent kernel	1
Sum kernel	$u + v$
Product kernel	uv
Brownian kernel	$\left(u^{1/3} + v^{1/3}\right)\left(u^{-1/3} + v^{-1/3}\right)$
EKE kernel	$\left(u^{1/3} + v^{1/3}\right)^2 \sqrt{u^{-1} + v^{-1}}$
Gravitational kernel	$\left(u^{1/3} + v^{1/3}\right)^2 \mid u^{1/6} - v^{1/6} \mid$

the agglomeration kernel can have a significant influence on the qualitative process behavior. A selection of suitable kernels has been evaluated and identified for a laboratory scale continuous fluidized bed spray agglomeration in Golovin et. al. [3, 4].

In order to obtain the birth rate of particles with volume v, Eq. (3) is integrated over the interval $[0, v]$:

$$B(t, v) = \frac{1}{2}\int_0^v \beta(t, u, v - u)n(t, u)n(t, v - u)\,\mathrm{d}u\,. \tag{4}$$

The death rate is defined analogously as

$$D(t, v) = \int_0^\infty \beta(t, v, u)n(t, v)n(t, u)\,\mathrm{d}u\,. \tag{5}$$

Finally, the agglomeration term is given as

$$\dot{n}_\mathrm{a}(t, v) = \frac{1}{2}\int_0^v \beta(t, u, v - u)n(t, u)n(t, v - u)\,\mathrm{d}u - \int_0^\infty \beta(t, v, u)n(t, v)n(t, u)\,\mathrm{d}u\,. \tag{6}$$

The particle feed is modeled as the product of the normalized number density distribution $q_{0,\mathrm{f}}(v)$ and the total number $N_\mathrm{f}(t)$ of added particles:

$$\dot{n}_\mathrm{f}(t, v) = N_\mathrm{f}(t)q_{0,\mathrm{f}}(v). \tag{7}$$

For the particle outlet it is assumed that particles exceeding a specific volume v_prod are removed from the process. Therefore, the separation function $T(v)$ is introduced. Since the separation is not ideal, $T(v)$ is modeled as a cumulative Gaussian function

$$T(v) = \int_0^v \frac{1}{\sqrt{2\pi\sigma^2}}\exp\left\{\frac{(s - v_\mathrm{prod})^2)}{\sigma^2}\right\}\mathrm{d}s\,, \tag{8}$$

where σ is a measure of the classification quality. The number density of removed particles is then defined as follows:

$$\dot{n}_\mathrm{o}(t, v) = K(t)T(v)n(t, v) \tag{9}$$

with removal rate $K(t)$. Inserting Eqs. (6), (7), and (9) in Eq. (1) yields the final process model:

$$\begin{aligned}\frac{\partial n(t,v)}{\partial t} =& \frac{1}{2}\int_0^v \beta(t,u,v-u)n(t,u)n(t,v-u)\,\mathrm{d}u - \int_0^\infty \beta(t,v,u)n(t,v)n(t,u)\,\mathrm{d}u \\ &+ N_\mathrm{f}(t)q_{0,\mathrm{f}}(v) - K(t)T(v)n(t,v).\end{aligned} \tag{10}$$

For general kernels $\beta(t,u,v)$ this PDE cannot be solved analytically. Thus, numerical solution techniques have to be used. In this contribution the cell-average method developed by Kumar et al. [6] is utilized for process simulation.

3 Control of Fluidized Bed Spray Agglomeration

This section is concerned with the derivation of stabilizing controllers for the FBSA. Therefore, the method of discrepancy based control is introduced in the first subsection and applied to the process in the following subsection. Furthermore, simulation studies are given and practical problems such as robustness with respect to parametric uncertainties are discussed.

3.1 Introduction to Discrepancy Based Control

A discrepancy $\rho(\varphi(.,t),t)$ is a generalized distance measure. It measures the distance between the process state $\varphi(.,t)$, i.e. a solution of the distributed parameter system, and the equilibrium φ_0. Here, it is of great importance that not all properties of a metric or norm have to be fulfilled. In the following, the main properties and facts on stability with respect to two discrepancies are stated in accordance to [7, 8, 13, 14].

Definition 1 Discrepancy A discrepancy is a real valued functional $\rho = \rho[\varphi(.,t),t]$ with the following properties

1. $\rho(\varphi,t) \geq 0$.
2. $\rho(0,t) = 0$.
3. for an arbitrary process $\varphi = \varphi(.,t)$ the real valued functional $\rho(\varphi(.,t),t)$ is continuous with respect to t.

In the context of stability with respect to two discrepancies besides the discrepancy $\rho(\varphi(.,t,t))$, measuring the distance between $\varphi(.,t)$ and the equilibrium φ_0, a second time independent discrepancy ρ_0 is used. It describes the distance between the initial state $\varphi(.,0)$ and the equilibrium φ_0. The two discrepancies ρ and ρ_0 have to satisfy, that $\rho(\varphi(.,t),t)$ is continuous at time $t = t_0$ with respect to ρ_0

at $\rho_0 = 0$, i.e. for every $\varepsilon > 0$ and $t_0 > 0$ there exists a $\delta(\varepsilon, t_0) > 0$, such that from $\rho_0 \leqslant \delta(\varepsilon, t_0)$ it follows that $\rho < \varepsilon$.

Definition 2 Stability with respect to two discrepancies ρ and ρ_0

The equilibrium $\varphi_0 = 0$ is stable in the sense of Lyapunov with respect to the two discrepancies ρ and ρ_0 for all $t \geq t_0$ if for every $\varepsilon > 0$ and $t_0 \geq 0$ there exists a $\delta = \delta(\varepsilon, t_0) > 0$ such that for every process $\varphi(., t)$ with $\rho_0 < \delta(\varepsilon, t_0)$ it follows that $\rho < \varepsilon$ for all $t \geq t_0$. If in addition $\lim_{t\to\infty} \rho = 0$, then the equilibrium φ_0 is called asymptotically stable in the sense of Lyapunov with respect to the two discrepancies ρ and ρ_0.

Based on the stated stability concept, i.e. stability with respect to two discrepancies, an according Lyapunov functional V can be introduced.

Definition 3 Positivity with respect to a discrepancy ρ.

The functional $V = V[\varphi, t]$ is called positive with respect to the discrepancy ρ, if $V \geq 0$ and $V[0, t] = 0$ for all φ with $\rho(\varphi, t) < \infty$.

Definition 4 Positive definiteness with respect to a discrepancy ρ.

The functional $V = V[\varphi, t]$ is positive definite with respect to a discrepancy ρ, if $V \geq 0$ and $V[0, t] = 0$ for all φ with $\rho(\varphi, t) < \infty$ and for every $\varepsilon > 0$ there exists a $\delta = \delta(\varepsilon) > 0$, such that $V \geq \delta(\varepsilon)$ for all φ with $\rho[\varphi, t] \geq \varepsilon$.

The following two theorems state the conditions for a function V guaranteeing (asymptotic) stability with respect to two discrepancies.

Theorem 1 ([14]) *The process φ with the equilibrium $\varphi_0 = 0$ is stable with respect to the two discrepancies ρ and ρ_0 if and only if there exists a functional $V = V[\varphi, t]$ positive definite with respect to the discrepancy ρ, continuous at time $t = t_0$ with respect to ρ_0 at $\rho_0 = 0$ and not increasing along the process φ, i.e. $\dot{V} \leq 0$.*

Theorem 2 ([14]) *The process φ with the equilibrium $\varphi_0 = 0$ is asymptotically stable with respect to the two discrepancies ρ and ρ_0 if and only if there exists a functional $V = V[\varphi, t]$ positive definite with respect to the discrepancy ρ, continuous at time $t = t_0$ with respect to ρ_0 at $\rho_0 = 0$ and not increasing along the process φ, i.e. $\dot{V} \leq 0$, with $\lim_{t\to\infty} V = 0$.*

As has been discussed in Palis and Kienle [12], stability with respect to two discrepancies can be interpreted as special output stability, where the discrepancy defines a virtual system output. Therefore, given a system, which is stable with respect to two discrepancies, stability of the full system state in terms of a norm is guaranteed if the zero dynamics are stable.

3.2 Application to Fluidized Bed Spray Agglomeration

In this section the discrepancy based control introduced above is applied to FBSA. Therefore, appropriate discrepancies have to be found. For a continuous granulation process, Palis and Kienle [10–12] showed, that the differences between desired and actual moments of the size distribution

$$\Delta\mu_i(t) = \int_0^\infty v^i\,(n_\mathrm{d}(v) - n(t,v))\,\mathrm{d}v\,, \tag{11}$$

where $N_\mathrm{d}(v)$ is the desired NDD, are a suitable choice. Here, the zeroth and the first moment represent the total particle number and the total particle volume, respectively. In this contribution the zeroth moment is chosen as control variable since it can be interpreted physically and its impact on the agglomeration term $\dot{n}_\mathrm{a}$ is of greater significance. In the following, two discrepancy based controllers are derived and evaluated. The first controller is a continuous controller guaranteeing exponential convergence of the control error. The second controller is a discrepancy based sliding mode controller. Both approaches are then compared.

3.2.1 Discrepancy Based Control

For the control of the zeroth moment, the control error is given as

$$e = \Delta\mu_0 = \int_0^\infty (n_d - n)\mathrm{d}v, \tag{12}$$

where n_d is the desired steady state. The according discrepancy and Lyapunov functional can hence be chosen as

$$\rho = \frac{1}{2}e^2, \tag{13}$$

$$V = \frac{1}{2}e^2. \tag{14}$$

Obviously, the error, the discrepancy, and the Lyapunov functional vanish not only at the desired steady state distribution but for all distributions with an equal zeroth moment. In order to derive a discrepancy based control law, the first order time derivative of the Lyapunov functional along the system trajectories is calculated

$$\dot{V} = e\dot{e} = -e\int_0^\infty \frac{\partial n}{\partial t}\mathrm{d}v = -e\left(\int_0^\infty \dot{n}_\mathrm{a} + n_\mathrm{f}\,\mathrm{d}v - K\int_0^\infty Tn\,\mathrm{d}v\right). \tag{15}$$

Here, the withdrawal rate K is the control handle. In order to achieve exponential convergence of the Lyapunof functional, K is chosen as

$$K = \frac{1}{\int_0^\infty Tn \, \mathrm{d}v} \left(-ce + \int_0^\infty \dot{n}_\mathrm{a} + n_\mathrm{f} \, \mathrm{d}v \right), \tag{16}$$

where c is a positive constant determining the convergence rate. This gives

$$\dot{V} = -2cV. \tag{17}$$

Applying the control law therefore exponentially stabilizes the agglomeration process with respect to the introduced discrepancy, i.e. the zeroth moment converges exponentially to the desired value. If the zero dynamics of the system are asymptotically stable in terms of a norm, stability of the distribution in terms of the same norm follows. Since a full analysis of the zero dynamics is usually not feasible for this type of PDE, a local stability analysis, i.e. using the linearization around the desired steady state of the discretized system, has been conducted showing that the associated transfer function does not possess zeros in the right half-plane. Thus, the zero dynamics are at least locally stable.

Furthermore, it should be mentioned that the denominator in Eq. (16) can vanish for some distributions leading to an undefined control law or take values close to zero leading to high controller gains. Due to the latter, K has to be bounded in practical applications.

To verify the designed control laws, the system was simulated numerically using the process parameters from Table 2. As agglomeration kernel the Brownian kernel was used. In Figs. 3 and 4 simulation results comparing open-loop and closed-loop operation are presented. It is shown that the moments as well as the L_2-norm of $(n_\mathrm{d}-n)$ converge in both cases. While the zeroth moment and the L_2-norm converge faster in closed-loop operation the total particle volume μ_1 does not. To achieve better performance with respect to this measure, a two-dimensional controller using the feed rate as another manipulated variable could be derived.

Table 2 Process parameters

Parameter	Value
N_f	380,000
β_0	1×10^{-10}
v_prod	0.9 mm
σ	0.3 mm
K_nom	0.0125 s^{-1}
K_max	1 s^{-1}
γ	1×10^{-3}
c	0.2

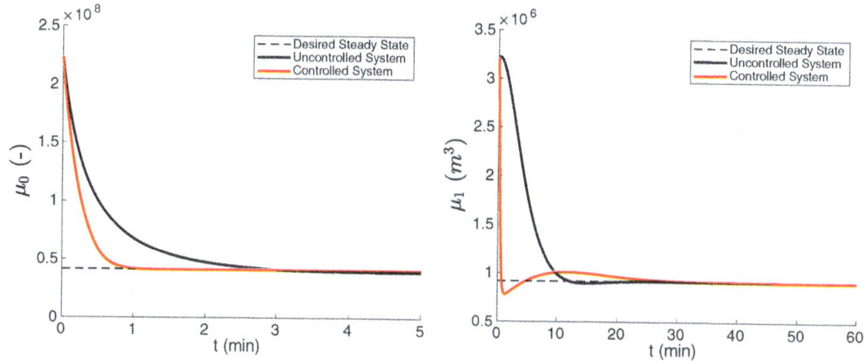

(a) Zeroth moment μ_0 (b) First moment μ_1

Fig. 3 Convergence of the zeroth (**a**) and the first moment (**b**) in closed-loop (red) and open-loop (black) operation

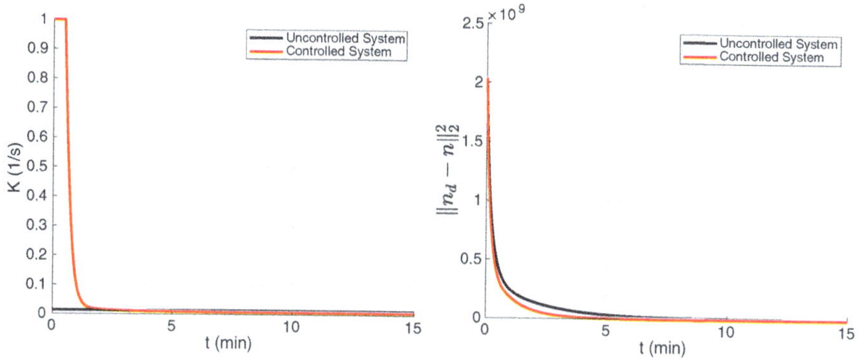

(a) (b)

Fig. 4 Manipulated variable K (left) and convergence in L_2-norm (right) in closed-loop (red) and open-loop (black) operation. (**a**) Withdrawal rate K. (**b**) L_2-norm

3.2.2 Discrepancy Based Sliding Mode Control

In this section a discrepancy based sliding mode controller is derived and tested using simulations. For the sake of simplicity and in order to improve comparability, the same discrepancy and Lyapunov functional as in the previous section are chosen. The time derivative of the Lyapunov functional along the system trajectory thus is

$$\dot{V} = -e\left(\int_0^{\infty} \dot{n}_{\mathrm{a}} + n_{\mathrm{f}}\,\mathrm{d}v - K\int_{v_{\mathrm{prod}}}^{\infty} n\,\mathrm{d}v\right). \tag{18}$$

Assuming there exists a maximum withdrawal rate $K_{\max} > 0$ with

$$\int_0^\infty \dot{n}_{\mathrm{a}} + n_{\mathrm{f}}\,\mathrm{d}v < K_{\max} \int_{v_{\mathrm{prod}}}^\infty n\,\mathrm{d}v \tag{19}$$

the following sliding mode control law

$$K = \begin{cases} 0 & \text{if } e \geq 0 \\ K_{\max} & \text{if } e < 0 \end{cases} \tag{20}$$

can be chosen, resulting in the required negative definiteness of the time derivative of the Lyapunov functional

$$\dot{V} \leq 0. \tag{21}$$

Therefore, the controller stabilizes the control variable.

Applying the derived sliding mode controller in the simulation setting results in the closed-loop behavior shown in Fig. 5. As can be seen on the left-hand side, by applying the discrepancy based sliding mode control law the zeroth moment converges. In contrast to the discrepancy based controller from the previous section, this happens in finite time. Shown on the right-hand side, the difference in the closed-loop convergence behavior of the first moment shown is however less significant. In Fig. 6 the phase portrait of the two moments (left) and the convergence in the L_2-norm (right) are depicted.

Besides the simple implementation of the discrepancy based sliding mode control law, a major advantage compared to the continuous control law is the robustness

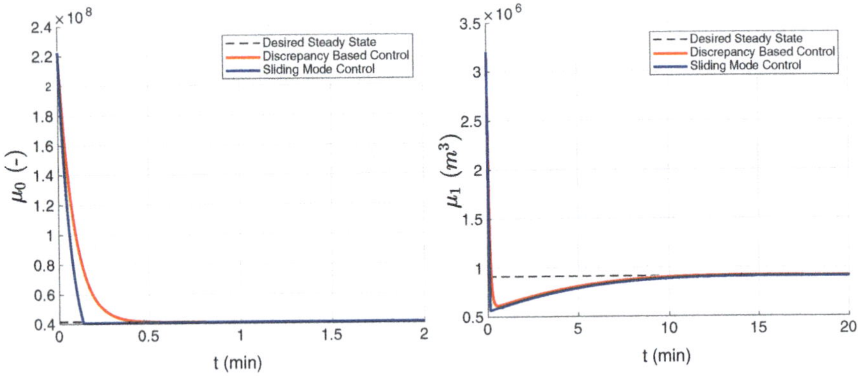

(a) (b)

Fig. 5 Convergence of the zeroth (left) and the first moment (right) with discrepancy based (red) and sliding mode(blue) control. (**a**) Zeroth moment μ_0. (**b**) First moment μ_1

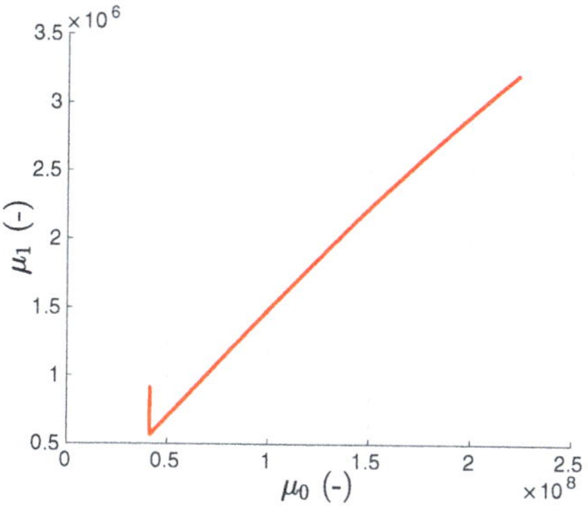

(a)

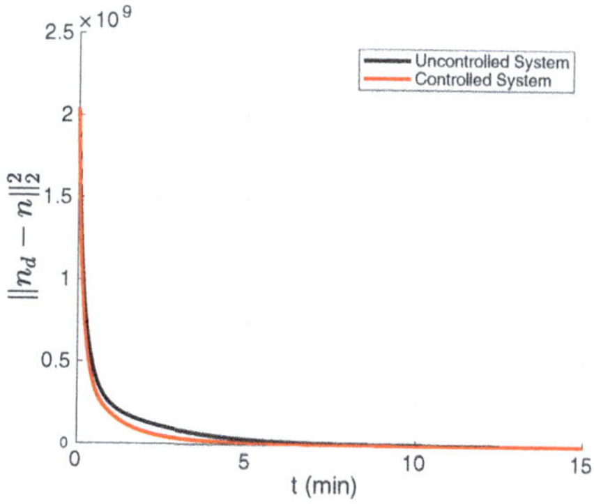

(b)

Fig. 6 Phase portrait (left) and convergence in L_2-norm (right) in closed-loop (red) and open-loop (black) operation. (**a**) Phase portrait in the μ_0–μ_1 plane. (**b**) L_2-norm

with respect to uncertainties in the model equations. This behavior is examined in the following section.

3.3 Robustness with Respect to Parametric Uncertainties

For practical implementation robustness of the control laws with respect to model uncertainties is an important feature. A typical parametric uncertainty for the given agglomeration process is the feed rate. While the proposed discrepancy based sliding mode control law does not depend on the feed and therefore possesses a natural robustness, the control law from Sect. 3.2.1 depends explicitly on the parameter N_{f}. Thus, stability is not guaranteed if the feed rate is disturbed. In order to compensate for this, the closed-loop control system can be augmented by a parameter estimator for the feed rate. Then the estimated feed rate is used to compute the according withdrawal rate. It has to be mentioned that the given parametric disturbance in the PDE changes the steady state distributions $n_s(v)$. Therefore, it is generally not possible to stabilize the desired distribution $n_d(v)$ under occurrence of disturbances even if the moments converge.

In order to derive a parameter estimation law the estimation error is defined using the unknown feed rate N_{f} and its estimate $\hat{N}_{\mathrm{f}}$ as follows:

$$\tilde{N}_{\mathrm{f}} = \hat{N}_{\mathrm{f}} - N_{\mathrm{f}}. \tag{22}$$

The certainty equivalence discrepancy based control law using the estimated feed rate $\hat{N}_{\mathrm{f}}$ is given as

$$K = \frac{1}{\int_0^\infty Tn\,\mathrm{d}v}\left(-ce + \int_0^\infty \dot{n}_{\mathrm{a}} + \hat{N}_{\mathrm{f}} q_{0,\mathrm{f}}\,\mathrm{d}v\right). \tag{23}$$

Following the well-known Lyapunov redesign approach, the Lyapunov functional is augmented with a term reflecting the estimation error

$$V = \frac{1}{2}e^2 + \frac{1}{2\gamma}\tilde{N}_{\mathrm{f}}^2, \tag{24}$$

where γ is a positive constant. Deriving the time derivative of the augmented Lyapunov functional along the closed-loop system trajectories, i.e. the controlled agglomeration process, yields

$$\dot{V} = e\dot{e} + \frac{1}{\gamma}\tilde{N}_{\mathrm{f}}\dot{\hat{N}}_{\mathrm{f}} = -ce^2 + \tilde{N}_{\mathrm{f}}\left(e + \frac{1}{\gamma}\dot{\hat{N}}_{\mathrm{f}}\right). \tag{25}$$

Here, the first term is as before negative definite in e. Due to the unknown sign of the estimation error $\tilde{N}_f$ the second term is indefinite. Therefore, an appropriate choice of the parameter update law $\dot{\hat{N}}_f$ is

$$\dot{\hat{N}}_{\mathrm{f}} = -\gamma e, \tag{26}$$

resulting in

$$\dot{V} = -ce^2. \tag{27}$$

Therefore, the error system is stable. To show asymptotic stability and thus convergence of the parameter estimate, LaSalles invariance principle can be used. Converging to and remaining at $\dot{V} = 0$ the error e has to vanish, i.e. $e = 0$. Here, the dynamics of e at $e = 0$ are given by

$$\dot{e} = -\int_0^\infty \dot{n}_{\mathrm{a}} + N_{\mathrm{f}} q_{0,\mathrm{f}} - KTn\,\mathrm{d}v\,. \tag{28}$$

After introducing the control law (23) and some simplifications this results in

$$\dot{e} = \tilde{N}_{\mathrm{f}}. \tag{29}$$

Therefore, the derivative of the control error vanishes only if the estimation error is also equal to zero. Thus, the estimated parameter converges asymptotically to the unknown parameter value.

In the following, simulation results for the system with a 50% disturbance in the feed rate, starting in the desired steady state, are presented. In Fig. 7 it can be seen that the zeroth moment μ_0 converges for both controllers, while the first moment is significantly smaller than desired. The sliding mode controller however has a better performance. The convergence of the parameter estimator and the (non-converging) L_2-norm are shown in Fig. 8.

Although it is, for the given configuration, generally not possible to achieve convergence of the L_2-norm in the presence of a non-vanishing disturbance due to the change of steady states, performance could be improved by using a different discrepancy, e.g. the average particle volume, which is the ratio between μ_1 and μ_0.

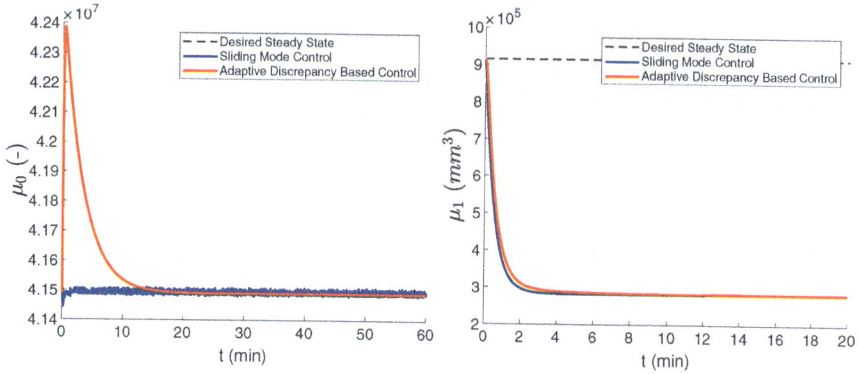

(a) (b)

Fig. 7 Convergence of the zeroth and first moment. (**a**) Zeroth moment μ_0. (**b**) First moment μ_1

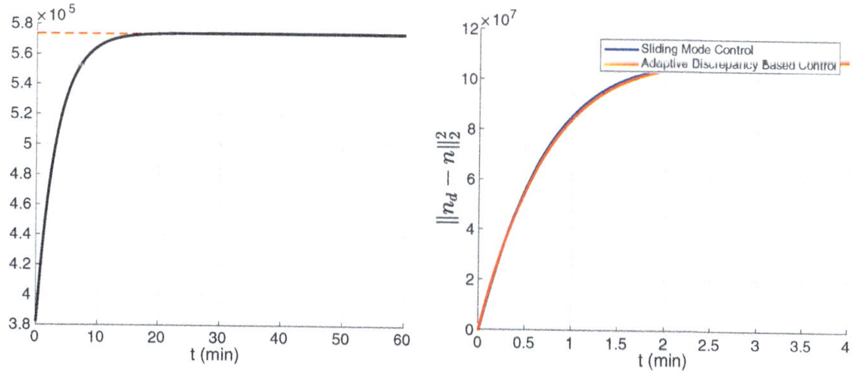

(a) Unknown parameter N_f (dotted red) and the parameter estimate $\hat{N}_f$ (black)

(b) L_2-norm of n

Fig. 8 Convergence of the parameter estimator (**a**) and L_2-norm of n (**b**)

4 Conclusion

In this contribution discrepancy based control for continuous fluidized bed spray agglomeration processes has been proposed. Here, two controller types, continuous and discontinuous, have been derived and analyzed. By design, both control laws guarantee stability with respect to the chosen discrepancy. For the given process configuration the zeroth moment of the number density distribution has been an appropriate choice. Furthermore, from a local stability analysis of the discretized zero dynamics and the simulation results it has been shown that the distributed state is also stabilized asymptotically in terms of a norm. General conditions for the stability of the zero dynamics could not be stated yet. Additionally, robustness with respect to parametric uncertainties was examined. Therefore, in order to guarantee asymptotic stability the continuous control law was augmented by a parameter estimator. Both the adaptive continuous controller and the sliding mode controller stabilize the chosen moment at the desired value.

References

1. Bück, A., Tsotsas, E.: Agglomeration. In: Caballero, B., Finglas, P.M., Toldrá, F. (eds.) Encyclopedia of Food and Health, pp. 73–81. Academic Press (2016). ISBN 9780123849533
2. Bück, A., Wegner, M., Neugebauer, C., Palis, S., Tsotsas, E.: Bifurcation analysis of process stability of continuous fluidized bed agglomeration with external product classification. In: 26th European Symposium on Computer Aided Process Engineering (Portoroz) (2016)
3. Golovin, I., Strenzke, G., Dürr, R., Palis, S., Bück, A., Tsotsas, E., Kienle, A.: Parameter identification for continuous fluidized bed spray agglomeration. Processes **6**, 246 (2018)
4. Golovin, I., Otto, E., Dürr, R., Palis, S., Kienle, A.: Lyapunov-based online parameter estimation in continuous fluidized bed spray agglomeration processes. In: 12th IFAC Symposium on Dynamics and Control of Process Systems - DYCOPS 2019 (Florianopolis) (2019)
5. Hulburt, H.M., Katz, S.: Some problems in particle technology: a statistical mechanical formulation. Chem. Eng. Sci. **19**, 555–574 (1964)
6. Kumar, J., Peglow, M., Warnecke, G., Heinrich, S.: The cell average technique for solving multi-dimensional aggregation population balance equations. Comput. Chem. Eng. **32**, 1810–1830 (2008)
7. Martynjuk, A., Gutovski, R.: Integral inequalities and stability of motion [in Russian]. Naukowa, Dumka (1979)
8. Movtschan, A.: Stability of processes with respect to two metrics [in Russian]. J. Appl. Math. Mech. **24**, 988–1001 (1960)
9. Palis, S.: Non-identifier-based adaptive control of continuous fluidized bed spray granulation. J. Process. Control **71**, 46–51 (2018)
10. Palis, S., Kienle, A.: Discrepancy based control of continuous crystallization. At-Automatisierungstechnik **60**, 145–154 (2012)
11. Palis, S., Kienle, A.: Stabilization of continuous fluidized bed spray granulation with external product classification. Chem. Eng. Sci. **70**, 200–209 (2012)
12. Palis, S., Kienle, A.: Discrepancy based control of particulate processes. J. Process. Control **24**, 33–46 (2014)

13. Sirasetdinov, T.: On the theory of stability of processes with distributed parameters [in Russian]. J. Appl. Math. Mech. **31**, 37–48 (1967)
14. Sirazetdinov, T.K.: Stability of Systems with Distributed Parameters [in Russian]. Nauka, Novosibirsk (1987)

On Polynomial Stability of Certain Class of $C0$-Semigroups

Grigory Sklyar and Piotr Polak

Abstract We consider linear differential equation given in Banach or Hilbert space and the semigroup corresponding to this equation. We give short overview of the results concerning asymptotic behavior of this semigroup. In particular, we discuss the existence of asymptotic stability, polynomial stability, and maximal asymptotics. We also give an analysis of asymptotic behavior of the semigroups generated by operators whose spectral family does not form Schauder basis.

1 Introduction

In this chapter, we give a survey of some results concerning asymptotic behavior of solutions of linear differential equations

$$\dot{x}(t) = Ax(t), \quad t \geq 0, \quad x(t) \in D(A) \subset X, \tag{1}$$

in abstract Banach or Hilbert space X, where densely defined operator $A : D(A) \to X$ generates strongly continuous semigroup of operators $\{T(t)\}_{t\geq 0}$. For any initial state $x \in X$, the function $T(t)x$ is said to be the solution of Eq. (1), and if initial state x belongs to the domain of generator $D(A)$, the function $T(t)x$ is differentiable and it is a classical solution of Eq. (1). In many cases, it is impossible to find semigroup $T(t)$ directly from the equation, i.e., it is impossible to solve Eq. (1) explicitly, which is why the studying of behavior of solutions via the properties of generator A is very important. In this chapter, we would like to summarize our last results concerning the study of behavior of solutions of linear differential equations by the location of the spectrum of generator A. It is known that the behavior of $\|T(t)\|$ cannot be described only by the location of the spectrum of generator A (which we will

G. Sklyar (✉) · P. Polak
Institute of Mathematics, University of Szczecin, Szczecin, Poland
e-mail: grigorij.sklyar@usz.edu.pl; piotr.polak@usz.edu.pl

G. Sklyar, A. Zuyev (eds.), *Stabilization of Distributed Parameter Systems: Design Methods and Applications*, SEMA SIMAI Springer Series 2,
https://doi.org/10.1007/978-3-030-61742-4_6

denote by $\sigma(A)$). In particular, the growth bound $\omega_0(T)$ may be strictly greater than the spectral bound $s(A) = \sup\{\mathrm{Re}\lambda, \lambda \in \sigma(A)\}$. We are mainly interested in how the spectrum of generator affects the behavior of semigroup or particular solutions and when this behavior can be described via the spectral properties of generator. For the case when $\omega_0(T) = 0$, the crucial question that can be asked in the context of behavior of the solutions is if the solutions are asymptotically stable or not. The answer was given by Sklyar–Shirman in 1982 [17] (for bounded generators) and later by Arendt–Batty [1] and Lyubich–Phong in 1988 [8] (not necessarily bounded generators) in the following theorem.

Theorem 1 ([1] and [8]) *Let A be the generator of a bounded C_0-semigroup $\{T(t)\}_{t\geq 0}$ on a Banach space X, and let $\sigma(A) \cap (i\mathbb{R})$ be at most countable. Then, the semigroup $\{T(t)\}_{t\geq 0}$ is strongly asymptotically stable, i.e.,*

$$\lim_{t\to+\infty} \|T(t)x\| = 0, \quad x \in X$$

if and only if the adjoint operator A^* has no pure imaginary eigenvalues. □

Of course, for $\omega_0(T) = 0$, the norm of semigroup $\|T(t)\|$ cannot tend to zero, and in the case, if $\|T(t)\|$ is bounded, Theorem 1 shows that the function $\|T(t)\|$ may not describe well the behavior of any solutions $T(t)x$. However, we cannot expect that there will be the other function, say $g(t)$, that would tend to zero and still be the upper bound of all solutions at least in the sense that

$$\|T(t)x\| \leq g(t) \cdot M_x, \quad t \geq 0, x \in X.$$

The existence of such function $g(t)$ would immediately imply (by Uniform Boundedness Principle) that $\|T(t)\| \to 0$, and as a consequence, $\omega_0(T) < 0$—a contradiction. Hence, there is no vanishing function $g(t)$ that would be the upper bound for the norm of all orbits $\|T(t)x\|$.

2 Asymptotic Behavior on Dense Subsets

If we restrict initial states x to the domain of generator—$D(A)$, then the following Batty's theorem points out the vanishing function that describes uniform stability of orbits in the unit ball in the domain space.

Theorem 2 ([3]) *Let $\{T(t)\}_{t\geq 0}$ be bounded C_0-semigroup on Banach space X with generator A. If, in addition, $\sigma(A) \cap (i\mathbb{R}) = \emptyset$, then*

$$\|T(t)A^{-1}\| \to 0, \qquad t \to +\infty. \tag{2}$$

Remark Condition (2) is equivalent to the existence of function $g(t) \to 0$ for $t \to +\infty$, namely $g(t) := \|T(t)A^{-1}\|$ such that

$$\|T(t)x\| = \mathcal{O}(g(t)), \quad t \to +\infty, \quad x \in D(A).$$

In the work [11], it was shown that Theorem 2 can be derived from Theorem 1, namely the following theorem holds.

Theorem 3 ([11]) *Let $\{T(t)\}_{t\geq 0}$ be a C_0-semigroup on Banach space X with generator A. Then, for any $\lambda \notin \sigma(A)$,*

$$\text{(a)} \quad \|T(t)(A-\lambda I)^{-1}\| \overset{t\to\infty}{\longrightarrow} 0 \quad \Longrightarrow \quad \sigma(A) \subset \{\operatorname{Re}\lambda < 0\}.$$

If, in addition, semigroup $T(t)$ is bounded, then

$$\text{(b)} \quad \|T(t)(A-\lambda I)^{-1}\| \overset{t\to\infty}{\longrightarrow} 0 \quad \Longleftarrow \quad \sigma(A) \subset \{\operatorname{Re}\lambda < 0\}.$$

Remark The part (a) of above theorem has been proved earlier for bounded semigroups by Batty and Duyckaerts in [4], and the part (b) is equivalent to Theorem 2. □

The function $\|T(t)A^{-1}\|$ gives the uniform upper bound of growth/decay of all orbits that start in the domain of generator. If the rate of decay of this function is polynomial (i.e., $t^{-\alpha}$ for some positive α), then the semigroup $T(t)$ is polynomially stable. The following detailed definition was given by Bátkai, Engel, Prüss, and Schnaubelt.

Definition ([2]) We call the semigroup (or corresponding equation) polynomially stable, if there exist positive constants M, α, and β, such that

$$\|T(t)x\| \leq Mt^{-\beta}\|x\|_{D(A^\alpha)}, \quad t > 0, x \in D(A^\alpha),$$

or equivalently

$$\|T(t)A^{-\alpha}\| \leq Mt^{-\beta}, \quad t \geq 0.$$

Borichev and Tomilov obtained the criterion of polynomial stability for bounded C_0-semigroups acting on Hilbert space, namely the following theorem holds.

Theorem 4 ([5]) *Let $\{T(t)\}_{t\geq 0}$ be a bounded C_0-semigroup on Hilbert space H with generator A satisfying $i\mathbb{R} \subset \rho(A)$. Then, for any fixed $\alpha > 0$, the following 3 conditions are equivalent:*

$$\|T(t)x\| \leq Mt^{-\frac{1}{\alpha}}\|x\|_{D(A)}, \quad t \to +\infty, x \in D(A), \quad \textit{for some } M > 0,$$

$$\|T(t)x\| \leq Mt^{-1}\|x\|_{D(A^\alpha)}, \quad t \to +\infty, x \in D(A^\alpha), \quad \textit{for some } M > 0,$$

$$\|R(A, is)\| = O(|s|^\alpha), \quad s \to \pm\infty. \tag{3}$$

In practice, finding the direct formula for resolvent operator is almost as difficult as finding the semigroup; hence, in general, the condition (3) is very hard to verify. However, Bátkai, Engel, Prüss, and Schnaubelt showed that similar condition implies some restriction on the location of the spectrum of generator.

Theorem 5 ([2]) *Let A generate bounded C_0-semigroup on Banach space X, and let $\sigma(A) \subset \mathbb{C}_-$. If*

$$\|R(A, is)A^{-\alpha}\| \leq C, \quad s \in \mathbb{R}, \tag{4}$$

then there exists $\delta > 0$ such that

$$|Im\lambda| \geq C|Re\lambda|^{-\frac{1}{\alpha}} \tag{5}$$

holds for all $\lambda \in \sigma(A) : |Re\lambda| < \delta$.

Remark In the work of Latushkin and Shvydkoy [7], it was shown that the conditions (3) and (4) on the behavior of the resolvent operator are equivalent. This means that in the case of a bounded semigroup on Hilbert space, the condition (5) on the location of the spectrum of generator is necessary for polynomial stability of semigroup, as the consequence of Theorems 4 and 5. □

In general, the condition on location of the spectrum of generator cannot be sufficient to determine the asymptotic behavior of semigroup or individual solutions or to determine the stability of semigroup. However, for some class of discrete type semigroups, we are able to prove that condition (5) is sufficient to polynomial stability. Let us consider the following class of generators:

(A1) $A : D(A) \subset H \to H$ generates C_0 group in Hilbert space H.
(A2) $\sigma^{(p)}(A) = \bigcup_{k\in\mathbb{Z}} \sigma_k$, such that

- (a) $\sigma_i \cap \sigma_j = \emptyset$ for $i \neq j$,
- (b) $\#\sigma_k \leq N, \quad k \in \mathbb{Z}$, and
- (c) $\inf\{|\lambda - \mu| : \lambda \in \sigma_i, \mu \in \sigma_j, i \neq j\} = d > 0$.

(A3) Linear span of generalized eigenvectors of operator A is dense in H.

Generators of above class may appear, for example, in the analysis of delay equations of neutral type. In this case, corresponding semigroups (groups) behave asymptotically like $t^{N-1}e^{\omega_0 t}$ (see [9, 15]). Besides, as it was shown in [18, 19], generalized eigenvectors of operator A (satisfying (A1)–(A-3)) constitute Riesz basis of finite-dimensional subspaces (see [14] for more details). The existence of Riesz basis was the key tool in the proof of the following theorem, that the spectral condition (5) is sufficient for polynomial stability:

Theorem 6 ([15]) *Let operator A satisfy assumptions (A1)–(A3). If $\sigma(A) \subset \mathbb{C}_-$, and for some constants $C, \alpha > 0$ holds*

$$|\mathrm{Im}\,\lambda| \geq C|\mathrm{Re}\lambda|^{-\frac{1}{\alpha}} : \lambda \in \sigma(\mathcal{A}).$$

Then, there exists constant $M > 0$ such that

$$\|T(t)x\| \leq Mt^{N-1-n\alpha}, \quad t > 1, \quad x \in D(\mathcal{A}^n).$$

The above theorem points out the dense set of initial states for which corresponding orbits are uniformly stable, while the whole semigroup is unbounded.

3 Maximal Asymptotics

The other aspect of asymptotic behavior of solutions is existence or lack of the solution that has maximal (minimal) rate of growth (decay). Such maximal rate of growth is called Maximal Asymptotics. This notion was provided first in [10] and formally defined as follows.

Definition ([10]) We say that equation $\dot{x} = Ax$ (or the semigroup $\{T(t), \ t \geq 0\}$) has a maximal asymptotics if there exists a real positive function $f(t), t \geq 0$, such that

(i) for any initial vector $x \in X$, the function $\frac{\|T(t)x\|}{f(t)}$ is bounded on $[0, +\infty)$, and
(ii) there exists at least one $x_0 \in X$ such that

$$\lim_{t \to +\infty} \frac{\|T(t)x_0\|}{f(t)} = 1.$$

If semigroup $T(t)$ acts on finite-dimensional space, then maximal asymptotics always exists and it is $t^{N-1}e^{\omega_0 t}$, where N is maximal algebraic multiplicity of eigenvalue with real part ω_0. For infinite-dimensional systems, it is much more complicated, and maximal asymptotics may not exist (see [10]). In particular,

Theorem 7 ([10]) *Assume that*

(i) $\sigma(A) \cap \{\lambda : Re\, \lambda = \omega_0\}$ is at most countable;
(ii) operator A^* does not possess eigenvalues with real part ω_0.

Then, equation $\dot{x} = Ax$ (the semigroup $\{T(t), \ t \geq 0\}$) does not have any maximal asymptotics. □

The lack of maximal asymptotics could mean that asymptotic behavior of "the fastest" solutions is not commensurate with $\|T(t)\|$; that is, the limits of $\|T(t)x\|/\|T(t)\|$ do not exist for some solutions or this could mean that all solutions grow much slower than $\|T(t)\|$, i.e., the limit $\|T(t)x\|/\|T(t)\|$ is zero for all solutions. At least for the case when the function $\|T(t)\|$ satisfies additional convexity condition, the lack of maximal asymptotics means that considered limits are 0 for all initial states; namely, we have the following theorem.

Theorem 8 ([10]) *Let the assumptions of Theorem 7 be satisfied (lack of maximal asymptotics), and let $f(t)$, $t \geq 0$ be a positive function such that*

(a) $\log f(t)$ *is concave,*
(b) for any $x \in X$, *the function* $\|T(t)x\|/f(t)$ *is bounded.* □

Then,

$$\lim_{t\to+\infty} \|T(t)x\|/f(t) = 0, \quad x \in X.$$

In this context, Theorem 1 may be written as follows: If bounded, a semigroup has generator A, which has at most countable number of eigenvalues on imaginary axis and the adjoint operator A^* has no eigenvalues on imaginary axis, then the semigroup does not have any maximal asymptotics, in particular, $\|T(t)\| \sim$ const is also not the one. Therefore, any solution $T(t)x$ grows slower than the constant function, which means that the solution must be asymptotically stable. Looking this way, Theorem 7 may be seen as a generalization of Theorem 1. Moreover, Theorem 7 may be applied in the case of unbounded semigroups.

If the generator of a semigroup has corresponding Riesz basis consisted of finite-dimensional subspaces, where the dimensions are uniformly bounded, then the following observation may be easily proved: if the function $f(t) := \|T(t)\|$ is not maximal asymptotics in all of the basis subspaces, then it cannot be the one in the whole space. The next theorem describes the similar situation without the assumption that subspaces constitute the Riesz basis

Theorem 9 ([16]) *Let* $\{T(t)\}_{t\geq 0}$ *be a* C_0*-semigroup of operators acting on Banach space* X, *with generator* A *and* $\omega_0(T) = 0$. *If*

(A) for any $\lambda \in \sigma(A) \cap (i\mathbb{R})$, *there exists a closed and bounded component of* $\sigma(A)$, *say* σ_λ, *containing* λ *(i.e.,* $\lambda \in \sigma_\lambda \subset \sigma(A)$*) and regular bounded curve* Γ_λ *enclosing* σ_λ, *such that* $\Gamma_\lambda \cap \sigma(A) = \emptyset$.
(B) for any $\lambda \in \sigma(A) \cap (i\mathbb{R})$ *and* $x \in X_\lambda$,

$$\lim_{t\to+\infty} \frac{\|T(t)x\|}{\|T(t)\|} \to 0,$$

where X_λ *is an image of the Riesz projection corresponding to the curve* Γ_λ.

Then, the semigroup $T(t)$ *has no maximal asymptotics.* □

Remark Theorem 9 stays true also without assumption that $\omega_0(T) = 0$; then, of course, one needs to study the spectrum near the line $\operatorname{Re}\lambda = \omega_0$. □

4 Asymptotic Behavior of a Certain Special Class of Semigroups

To illustrate Theorem 9, we define the operator A with the spectral family of eigenvectors, which does not form Riesz (or even Schauder) basis. The detailed construction of such operator and the analysis of its properties were first performed in [12, 13]; we recall the sketch of the construction. We start from Hilbert space $(X', \|\cdot\|)$ with orthonormal basis $(e_n)_{n\in\mathbb{N}}$. We define operator $A' : X' \to X'$ as follows:

$$A'e_n := i \ln n \, e_n, \quad n \in \mathbb{N}, \tag{6}$$

with the domain

$$D(A') := \left\{ x = \sum_{n=1}^{\infty} c_n e_n : \sum_{n=1}^{\infty} |c_n \ln n|^2 < \infty \right\}. \tag{7}$$

Of course, the operator $A' : X' \to X'$ possesses corresponding Riesz basis. Next, we define new norm in the space X' as follows:

$$\|x\|_1^2 := \sum_{n=1}^{\infty} |\Delta c_n|^2 := \sum_{n=1}^{\infty} (c_n - c_{n-1})^2,$$

for any $x = \sum_{n=1}^{\infty} c_n e_n$, where $c_0 := 0$. Let $(X, \|\cdot\|_1)$ be the completion of space $(X', \|\cdot\|_1)$. Space X may be seen as a space of formal sequences $(f)\sum c_n e_n$ such that $\sum(c_n - c_{n-1})^2$ converges. It occurs (see [12, 13]) that the sequence $(e_n)_{n\in\mathbb{N}} \subset X$ is a complete system, but it is not even Schauder basis. We consider operator $A : D(A) \subset X \to X$ given by the same formula that A' but acting in completed space X, where

$$D(A) := \left\{ (f)\sum_{n=1}^{\infty} c_n e_n : \sum_{n=1}^{\infty} |\Delta(c_n \ln n)|^2 < \infty \right\}. \tag{8}$$

It is proved that A is a generator of unbounded C_0-group in X, say $T(t)$, which grows like t, i.e., there exist constants $m, M > 0$ such that

$$m|t| \le \|T(t)\| \le M(|t| + 1), \quad t \in \mathbb{R}. \tag{9}$$

The key tool in proofs of these facts was the following Hardy's inequality

Theorem 10 ([6]) *Let $p > 1$ and (a_n) be a real positive sequence, and then*

$$\sum_{n=1}^{\infty} \left(\frac{a_1 + \ldots + a_n}{n} \right)^p \le \left(\frac{p}{p-1} \right)^p \sum_{n=1}^{\infty} a_n^p,$$

or equivalently, for increasing sequence b_n,

$$\sum_{n=1}^{\infty}\left(\frac{b_n}{n}\right)^p \le \left(\frac{p}{p-1}\right)^p \sum_{n=1}^{\infty}(\Delta b_n)^p.$$

Remark It is easy to see that for even p, the requirement that the sequence (a_n) should be positive or (b_n) increasing is not necessary. Moreover, for $p = 2$ and any complex sequence c_n, there is

$$\sum_{n=1}^{\infty}\left|\frac{c_n}{n}\right|^2 \le 4\sum_{n=1}^{\infty}|\Delta c_n|^2. \tag{10}$$

The last inequality follows from the previous one for $b_n = |c_n|$ and the triangle inequality $|\Delta b_n| \le |\Delta c_n|$.

The estimate (9) means that the semigroup $T(t)$ is unbounded, $\omega_0(T) = 0$, and due to Theorem 9, it may be checked easily that the semigroup does not possess any maximal asymptotics, but even that the semigroup cannot be asymptotically stable or polynomially stable because its generator has pure imaginary eigenvalues (see Theorem 3). We may check the asymptotic behavior of solutions starting in $D(A^k)$ for any $k \in \mathbb{N}$ directly.

Statement Let $A : D(A) \subset X \to X$ be the operator defined by (6),with domain (8), and let $k \in \mathbb{N}$ be fixed. Then, there exist constant $M > 0$ and time $t_0 > 0$ such that

$$\|T(t)x\|_1 \le M\frac{t}{\ln^k t}\|x\|_{D(A^k)}, \quad x \in D(A^k), \quad t > t_0,$$

where $\|\cdot\|_{D(A^k)}$ is an appropriate graph norm, i.e., $\|x\|_{D(A^k)} := \|x\|_1 + \|A^k x\|_1$. □

Proof Let us take any $x \in D(A^k)$. This means

$$x = (f)\sum_{n=1}^{\infty} c_n e_n : \quad \sum_{n=1}^{\infty}|\Delta(c_n \ln^k n)|^2 < \infty.$$

Consider $\|T(t)x\|_1$:

$$\|T(t)x\|_1^2 = \sum_{n=1}^{\infty}\left|\Delta(c_n e^{it\ln n})\right|^2 \le 2\sum_{n=1}^{\infty}\left(|\Delta c_n|^2 + |c_{n-1}|^2\left|\Delta(e^{it\ln n})\right|^2\right),$$

where we have used the property $\Delta(a_n \cdot b_n) = b_n \cdot \Delta a_n + a_{n-1} \cdot \Delta b_n$. Taking into account that $c_0 = 0$, we get

$$\|T(t)x\|_1^2 \leq 2\sum_{n=1}^{\infty} |\Delta c_n|^2 + 2\sum_{n=1}^{\infty} |c_n|^2 \left|\Delta(e^{it\ln(n+1)})\right|^2. \tag{11}$$

The first sum may be estimated easily as follows:

$$\sum_{n=1}^{\infty} |\Delta c_n|^2 = \|x\|_1^2 \leq \|x\|_{D(A^k)}^2 \leq \left(\frac{t}{\ln^k t}\right)^2 \|x\|_{D(A^k)}^2, \tag{12}$$

where the last inequality holds for $t > t_0$. The second sum in (11) needs more effort, namely

$$\sum_{n=1}^{\infty} |c_n|^2 \left|\Delta(e^{it\ln(n+1)})\right|^2 = \sum_{n=1}^{\infty} |c_n|^2 \left|e^{it\ln\left(1+\frac{1}{n}\right)} - 1\right|^2. \tag{13}$$

Let us split the above sum for $n < t$ and for $n \geq t$. We estimate the first part

$$\sum_{n<t} |c_n|^2 \left|e^{it\ln\left(1+\frac{1}{n}\right)} - 1\right|^2 \leq 4\sum_{n<t} |c_n|^2 \leq 4|c_1|^2 + 4\sum_{2\leq n<t} \left|\frac{c_n \ln^k n}{n}\right|^2 \cdot \left(\frac{n}{\ln^k n}\right)^2.$$

For any $k \in \mathbb{N}$, the function $\frac{t}{\ln^k t}$ increases monotonically to $+\infty$ for $t > \ln k$. Thus, for fixed $k \in \mathbb{N}$, there exists t_0 such that $\frac{n}{\ln^k n} \leq \frac{t}{\ln^k t}$ for $n \in [2, t)$ and $t > t_0$. Hence, for $t > t_0$, we obtain

$$\begin{aligned}\sum_{n<t} |c_n|^2 \left|e^{it\ln\left(1+\frac{1}{n}\right)} - 1\right|^2 &\leq 4|c_1|^2 + 4\left(\frac{t}{\ln^k t}\right)^2 \sum_{1<n<t} \left|\frac{c_n \ln^k n}{n}\right|^2 \\ &\leq 4|\Delta c_1|^2 + 4\left(\frac{t}{\ln^k t}\right)^2 \sum_{n=1}^{\infty} \left|\frac{c_n \ln^k n}{n}\right|^2.\end{aligned}$$

Using Hardy's inequality (10), we obtain

$$\sum_{n<t} |c_n|^2 \left|e^{it\ln\left(1+\frac{1}{n}\right)} - 1\right|^2 \leq 4\|x\|_1^2 + 16\left(\frac{t}{\ln^k t}\right)^2 \|A^k x\|_1^2 \leq 16\left(\frac{t}{\ln^k t}\right)^2 \|x\|_{D(A^k)}^2. \tag{14}$$

For the second part of the sum (13), for $n > t$, we proceed as follows: due to the inequality $t\ln\left(1+\frac{1}{n}\right) \le \frac{t}{n} < 1$ for $n > t$, we also have

$$\left|e^{it\ln\left(1+\frac{1}{n}\right)} - 1\right| \le \left|e^{i\frac{t}{n}} - 1\right|. \tag{15}$$

The left-hand side of above inequality is the length of base of isosceles triangle with vertices: $0, 1, e^{it\ln\left(1+\frac{1}{n}\right)}$ and angle $t\ln\left(1+\frac{1}{n}\right) < 1$, while the right-hand side is the base of isosceles triangle with complex vertices: $0; 1; e^{i\frac{t}{n}}$ and bigger angle $\frac{t}{n}$; hence, the above inequality follows from cosine rule. Inequality (15) in common with Hardy's inequality (10) allows us to obtain the following estimation:

$$\begin{aligned}
&\sum_{n>t} |c_n|^2 \left|e^{it\ln\left(1+\frac{1}{n}\right)} - 1\right|^2 \le \sum_{n>t} |c_n|^2 \left|e^{i\frac{t}{n}} - 1\right|^2 \le \\
&\le t^2 \sum_{n>t} \left|\frac{c_n \ln^k n}{n}\right|^2 \left|\frac{e^{i\frac{t}{n}} - 1}{\frac{t}{n}}\right|^2 \cdot \frac{1}{(\ln^k n)^2} \le M_k^2 t^2 \sum_{n>t} \left|\frac{c_n \ln^k n}{n}\right|^2 \frac{1}{(\ln^k t)^2} \\
&\le M_k^2 \left(\frac{t}{\ln^k t}\right)^2 \sum_{n=1}^{\infty} \left|\frac{c_n \ln^k n}{n}\right|^2 \le 4M_k^2 \left(\frac{t}{\ln^k t}\right)^2 \sum_{n=1}^{\infty} \left|\Delta(c_n \ln^k n)\right|^2 \\
&= 4M_k^2 \left(\frac{t}{\ln^k t}\right)^2 \left\|A^k x\right\|_1^2 \le 4M_k^2 \left(\frac{t}{\ln^k t}\right)^2 \|x\|_{D(A^k)}^2,
\end{aligned}$$

where we have used the fact that there exists constant $M_k \ge 1$ such that $\left|\frac{e^{ix}-1}{x}\right| \le M_k$, $x \in (0; +\infty)$. Collecting (11)–(14) and the above, we get

$$\|T(t)x\|_1 \le 4M_k \frac{t}{\ln^k t} \|x\|_{D(A^k)}, \quad x \in D(A^k), \quad t > t_0. \qquad \square$$

As we can see, taking initial states from the domain of generator in this particular case may decrease the asymptotic growth of solution only by logarithmic factor. To obtain better reduction of growth rate, we may try to pull the spectrum slightly to the left half-plane. Let us consider new operator $A'_\alpha : X' \to X', \alpha > 0$ whose spectrum is approaching imaginary axis sufficiently slow:

$$A'_\alpha e_n := \lambda_n e_n := (i\ln n - \frac{1}{\ln^\alpha n}) e_n, \quad n \in \mathbb{N}, \tag{16}$$

with the domain

$$D(A'_\alpha) := \{x = \sum_{n=1}^{\infty} c_n e_n : \sum_{n=1}^{\infty} |c_n \lambda_n|^2 < \infty\}. \tag{17}$$

Similarly like for A' (see Theorem 6 in [13]), it can be shown that each A'_α defines operator $A_\alpha : D(A_\alpha) \subset X \to X$ and A_α generates C_0-semigroup $\{T_\alpha(t)\}_{t\geq 0}$ on X given by

$$T_\alpha(t)x = (f)\sum_{n=1}^{\infty} c_n e^{\lambda_n t} e_n, \quad x = (f)\sum_{n=1}^{\infty} c_n e_n \in X.$$

Operators A_α do not satisfy assumption (A2)(c), so we cannot use Theorem 6 to estimate the behavior of solutions, nevertheless we can check this behavior directly.

Statement Let A_α be the operator defined by (16) and (17), and let $k \in \mathbb{N}$, $\alpha > 0$ be fixed. There exist constant $M > 0$ and time $t_0 > 0$ such that

$$\|T(t)x\|_1 \leq Mt^{1-\frac{k}{\alpha}} \|x\|_{D(A^k)}, \quad x \in D(A^k), \quad t > t_0,$$

where $\|\cdot\|_{D(A^k)}$ is an appropriate graph norm, i.e., $\|x\|_{D(A^k)} := \|x\|_1 + \|A^k x\|_1$.

Proof For any $x \in D(A^k)$, we will estimate $\|T(t)x\|_1$:

$$\|T(t)x\|_1^2 = \sum_{n=1}^{\infty} \left|\Delta(c_n e^{t\lambda_n})\right|^2 \tag{18}$$

$$\leq 2\sum_{n=1}^{\infty} \left(|\Delta c_n|^2 e^{2t\operatorname{Re}\lambda_n} + |c_{n-1}|^2 \left|\Delta(e^{t\lambda_n})\right|^2\right) \tag{19}$$

$$= 2\sum_{n=1}^{\infty} |\Delta c_n|^2 e^{2t\operatorname{Re}\lambda_n} + 2\sum_{n=1}^{\infty} |c_{n-1}|^2 \left|\Delta(e^{t\lambda_n})\right|^2. \tag{20}$$

We will show that both sums may be estimated by $Mt^{2-\frac{2k}{\alpha}} \|x\|^2_{D(A^k)}$, which will end the proof. The first sum may be represented as follows:

$$\sum_{n=1}^{\infty} |\Delta c_n|^2 e^{2t\operatorname{Re}\lambda_n} = \sum_{n=1}^{\infty} \left|\Delta(c_n)\lambda_n^k\right|^2 \left(\frac{t}{\ln^\alpha n}\right)^{\frac{2k}{\alpha}} e^{-\frac{2t}{\ln^\alpha n}} \left(\frac{\ln^\alpha n}{t}\right)^{\frac{2k}{\alpha}} \frac{1}{|\lambda_n|^{2k}}.$$

Taking into account that the function $\left(\frac{t}{\ln^\alpha n}\right)^{\frac{2k}{\alpha}} e^{-\frac{2t}{\ln^\alpha n}}$ is bounded for $t > 0, n > 1$ by the constant M_k and $|\lambda_n|^{2k} \geq (\operatorname{Im} \lambda_n)^{2k} = \ln^{2k} n$, we obtain

$$\sum_{n=1}^{\infty} |\Delta c_n|^2 e^{2t \operatorname{Re} \lambda_n} \leq M_k t^{-\frac{2k}{\alpha}} \sum_{n=1}^{\infty} \left| \Delta(c_n) \lambda_n^k \right|^2. \tag{21}$$

Notice that $\Delta(c_n \lambda_n^k) = \Delta(c_n)\lambda_n^k + c_{n-1}\Delta(\lambda_n^k)$ implies $|\Delta(c_n) \cdot \lambda_n^k| \leq |\Delta(c_n \lambda_n^k)| + |c_{n-1}\Delta(\lambda_n^k)|$, and next $\Delta(\lambda_n^k)$ may be estimated using the mean value theorem. Therefore,

$$\sum_{n=1}^{\infty} |\Delta c_n|^2 e^{2t \operatorname{Re} \lambda_n} \leq M_k t^{-\frac{2k}{\alpha}} \left(\|A^k x\|_1^2 + \sum_{n=1}^{\infty} \left| \frac{c_{n-1}}{\varepsilon(n)} \lambda_{\varepsilon(n)}^{k-1} \left(ik + \frac{\alpha k}{\ln^{\alpha+1} \varepsilon(n)} \right) \right|^2 \right),$$

where $\varepsilon(n) \in (n-1, n), n \in \mathbb{N}$. The sequence $ik + \frac{\alpha k}{\ln^{\alpha+1} \varepsilon(n)}$ is bounded, and thus for new constant M_k, we have

$$\sum_{n=1}^{\infty} |\Delta c_n|^2 e^{2t \operatorname{Re} \lambda_n} \leq M_k t^{-\frac{2k}{\alpha}} \left(\|A^k x\|_1^2 + \sum_{n=1}^{\infty} \left| \frac{c_{n-1} \lambda_{n-1}^{k-1}}{n} \right|^2 \cdot \left| \frac{\lambda_{\varepsilon(n)}^{k-1}}{\lambda_{n-1}^{k-1}} \frac{n}{\varepsilon(n)} \right|^2 \right).$$

Once again, the sequence $\left| \frac{\lambda_{\varepsilon(n)}^{k-1}}{\lambda_{n-1}^{k-1}} \frac{n}{\varepsilon(n)} \right|^2$ is bounded, and by Hardy's inequality, we get

$$\begin{aligned} \sum_{n=1}^{\infty} |\Delta c_n|^2 e^{2t \operatorname{Re} \lambda_n} &\leq M_k t^{-\frac{2k}{\alpha}} \left(\|A^k x\|_1^2 + 4 \sum_{n=1}^{\infty} \left| \Delta(c_{n-1} \lambda_{n-1}^{k-1}) \right|^2 \right) \\ &\leq 4 M_k t^{-\frac{2k}{\alpha}} \left(\|A^k x\|_1^2 + \|A^{k-1} x\|_1^2 \right). \end{aligned}$$

Since $\|A^{k-1}x\|_1 \leq \|R(A, 0)\| \cdot \|A^{k-1}x\|_1$, we may write

$$\sum_{n=1}^{\infty} |\Delta c_n|^2 e^{2t \operatorname{Re} \lambda_n} \leq M_k t^{-\frac{2k}{\alpha}} \|A^k x\|_1^2 \leq M_k t^{-\frac{2k}{\alpha}} \|x\|_{D(A^k)}^2, \tag{22}$$

where $M_k > 0$ is a new constant. For the second sum, we do the similar operation, namely

$$\sum_{n=1}^{\infty} |c_{n-1}|^2 \left|\Delta(e^{t\lambda_n})\right|^2 = t^{2-\frac{2k}{\alpha}} \sum_{n=1}^{\infty} \left|\frac{c_{n-1} \ln^k(n-1)}{n}\right|^2 \left(\frac{t}{\ln^\alpha \varepsilon(n)}\right)^{\frac{2k}{\alpha}} e^{-\frac{2t}{\ln^\alpha \varepsilon(n)}}$$

$$\cdot \left|i + \frac{\alpha}{\ln^{\alpha+1} \varepsilon(n)}\right|^2 \cdot \left|\frac{n}{\varepsilon(n)} \frac{\ln^k \varepsilon(n)}{\ln^k(n-1)}\right|^2$$

$$\leq M_k t^{2-\frac{2k}{\alpha}} \|x\|^2_{D(A^k)}.$$

Using the above and (22) in (20), we get the assertion. □

References

1. Arendt, W., Batty, C.J.K.: Tauberian theorems for one-parameter semigroups. Trans. Amer. Math. Soc. **306**, 837–852 (1988)
2. Bátkai, A., Engel, K.-J., Pruss, J., Schnaubelt, R.: Polynomial stability of operator semigroups. Math. Nachr. **279**, 1425–1440 (2006)
3. Batty, C.J.K.: Tauberian theorems for the Laplace-Stieltjes transform. Trans. Amer. Math. Soc. **322**, 783–804 (1990)
4. Batty, Ch., Duyckaerts, T.: Non-uniform stability for bounded semi-groups on Banach spaces. J. Evol. Equ. **8**, 765–780 (2008)
5. Borichev, A., Tomilov, Y.: Optimal polynomial decay of functions and operator semigroups. Math. Ann. **347**, 455–478 (2010)
6. Hardy, G.H., Littlewood, J.E., Polya, G.: Inequalities. Cambridge University Press, Cambridge (1964)
7. Latushkin, Y., Shvydkoy, R.: Hyperbolicity of semigroups and Fourier multipliers. In: Systems, Approximation, Singular Integral Operators, and Related Topics (Bordeaux, 2000). Operator Theory: Advances and Applications, vol. 129, pp. 341–363. Birkhäuser, Basel (2001)
8. Lyubich, Yu.I., Vu, Q.P.: Asymptotic stability of linear differential equations on Banach spaces. Studia Math. **88**, 37–42 (1988)
9. Rabah, R., Sklyar, G.M., Rezounenko, A.V.: Stability analysis of neutral type systems in Hilbert space. J. Differ. Equ. **214**, 391–428 (2005)
10. Sklyar, G.M.: On the maximal asymptotics for linear differential equations in Banach spaces. Taiwanese J. Math. **14**, 2203–2217 (2010)
11. Sklyar, G.M.: On the decay of bounded semigroup on the domain of its generator. Vietnam J. Math. **43**, 207–213 (2015)
12. Sklyar, G.M., Marchenko, V.: Hardy inequality and the construction of the generator of a C0-group with eigenvectors not forming a basis. Dopov. Nac. akad. nauk Ukr. **9**, 13–17 (2015)
13. Sklyar, G.M., Marchenko, V.: Hardy inequality and the construction of infinitesimal operators with non-basis family of eigenvectors. J. Funct. Anal. **272**, 1017–1043 (2017)
14. Sklyar, G.M., Marchenko, V.: Resolvent of the generator of the C_0-group with non-basis family of eigenvectors and sharpness of the XYZ theorem. J. Spect. Theory Accepted. arXiv:1809.03079
15. Sklyar, G.M., Polak, P.: On asymptotic estimation of a discrete type C_0-semigroups on dense sets: application to neutral type systems. Appl. Math. Optim. **75**, 175–192 (2017)

16. Sklyar, G.M., Polak, P.: Notes on the asymptotic properties of some class of unbounded strongly continuous semigroups. J. Math. Phys. Anal. Geo. **75**, 412–424 (2019)
17. Sklyar, G.M., Shirman, V.: On asymptotic stability of linear differential equation in Banach space. Teoria Funk. Funkt. Anal. Prilozh. **37**, 127–132 (1982) (in Russian)
18. Xu, G.Q., Yung, S.P.: The expansion of a semigroup and a Riesz basis criterion. J. Differ. Equ. **210**, 1–24 (2005)
19. Zwart, H.: Riesz basis for strongly continuous groups. J. Diff. Eq. **249**, 2397–2408 (2010)

Existence of Optimal Stability Margin for Weakly Damped Beams

Jarosław Woźniak and Mateusz Firkowski

Abstract We analyze stability of a particular model of vibrations of Timoshenko beams with a weak (distributed) damping connected to rotations of cross-sections of the beam, of deflections of the center line of the beam, and of both. In one of the cases considered, for some values of physical parameters of the beam the optimal stability margin phenomenon may be observed, which means that under some conditions there exists an optimal value of a damping coefficient, that is a coefficient that guarantees the fastest possible decay of norms of solutions of the system.

1 Introduction

Stability and stabilizability analysis are parts of mathematical control theory which are widely used in applied sciences such as engineering (e.g., automation and robotics [12, 15]), medicine (e.g., blood biochemistry [8, 13]), biomathematics (e.g., population analysis [3, 4, 6]), etc. One of areas still under active research is analysis of stability of systems with damping [2, 7, 10]. Here we focus on the problem of describing the stability margin of the system of vibrating Timoshenko beams, which is a special case of a general second order system

$$\begin{gathered} \ddot{z}(t) + B\dot{z}(t) + Az(t) = Cu(t), \; t \geq 0, \; z(0) = z_0, \; \dot{z}(0) = z_1, \\ y = Dz(t), \end{gathered} \tag{1}$$

J. Woźniak (✉)
Institute of Mathematics, University of Szczecin, Szczecin, Poland
e-mail: wozniak@univ.szczecin.pl

M. Firkowski
School of Mathematics, West Pomeranian University of Technology Szczecin, Szczecin, Poland
e-mail: mateusz.firkowski@zut.edu.pl

G. Sklyar, A. Zuyev (eds.), *Stabilization of Distributed Parameter Systems: Design Methods and Applications*, SEMA SIMAI Springer Series 2,
https://doi.org/10.1007/978-3-030-61742-4_7

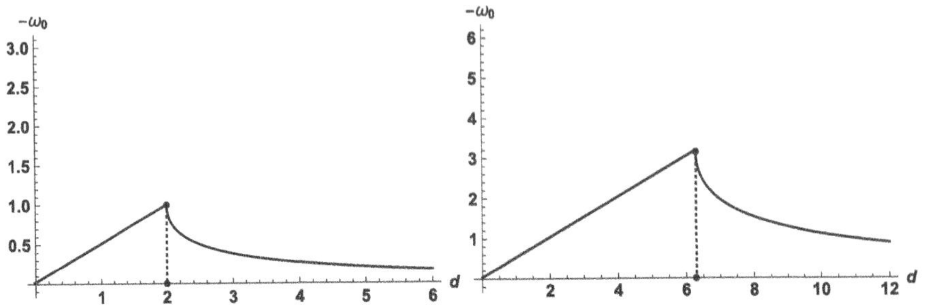

Fig. 1 Stability margin of damped spring–mass system (left) and damped string (right)

defined in suitable Hilbert spaces, and our considerations are based on finding decay rate of the system which in turn in many situations can be found by means of spectral analysis.

Let us describe this problem in more details.

Definition 1 (See [1]) A C_0-semigroup, $T(t)$, on a Hilbert space $\mathcal{H}$, generated by operator $\mathcal{A}$, connected with (1), is *exponentially stable* if there exist positive constants M and ω such that

$$\|T(t)\| \leq Me^{-\omega t} \text{ for } t \geq 0.$$

The constant ω is called the *decay rate*, and the supremum over all possible values of ω is the *stability margin* of $T(t)$.

Stability margin measures robustness of the system, i.e., possibility of change of physical parameters preserving stability properties.

Analyzing different vibrating systems such as vibrating string or spring–mass system with friction of movement (see Fig. 1) one can observe under-, over-, and critical damping effect. It appears that in those systems there exists a critical damping coefficient which in turn can be interpreted as an optimal stability margin phenomenon. Namely, we will consider the following

Definition 2 The damping operator B_0 will be called *optimal*, if system (1) with $B = B_0$ admits the fastest possible energy dissipation, that is if B_0 maximizes the stability margin of the system.

The analysis considering decay rate or stability margin, including search of optimal stability margin, realizes heavily on properly conducted spectral analysis of unbounded operators appearing in (1). One must be careful though as the location of spectrum alone may but does not have to be directly connected to decay rate of the system in question, as explained in detail in Sect. 3.

In our previous works [19–22] we observed that in some cases of weakly damped vibrating Timoshenko beams there is no stability margin at all, in some cases one cannot distinguish the critical case, and in some cases there exists optimal stability margin we were looking for. In this paper we summarize and re-interpret our previous results and state some new open questions.

2 Weakly Damped Rotating Timoshenko Beams

We consider a general model of the rotation of a Timoshenko beam (see [5]) in horizontal plane whose left end is rigidly clamped into the disk of a driving motor with similar type of damping like in [19–22]. For simplicity, the model derivation will be presented in one case only. If $w(x,t)$ denotes the deflection of the center line of the beam at the location $x \in [0,1]$ (the length of beam is assumed to be 1) and the time $t \geq 0$, and $\xi(x,t)$—the rotation angle of the cross section area at x and t (see Fig. 2) and if we assume the rotation to be slow, w and ξ are governed by two partial differential equations

$$\begin{cases} -\tilde{\rho}A\ddot{w}(x,t) - \tilde{\rho}A(r+x)\ddot{\theta}(t) + K(w''(x,t)+\xi'(x,t)) = 0, \\ -\tilde{\rho}I(\ddot{\xi}(x,t)-\ddot{\theta}(t)) + EI\xi''(x,t) - K(w'(x,t)+\xi(x,t)) - \tilde{\nu}^2\dot{\xi}(x,t) = 0 \end{cases} \quad (2)$$

for $x \in (0,1)$ and $t > 0$, where $\dot{w} = w_t$, $w' = w_x$, $\dot{\xi} = \xi_t$ and $\xi' = \xi_x$. By $r > 0$ we denote the radius of the disk and let $\theta = \theta(t)$ be the rotation angle as a function of the time $t \geq 0$. To a (uniform) cross section of the beam at point x, with $0 \leq x \leq 1$, we assign the following: K—shear stiffness, $\tilde{\rho}$—(constant) density of the material of the beam, E—Young's modulus, A—cross section area, I—area moment of inertia, $\tilde{\nu}$—damping coefficient.

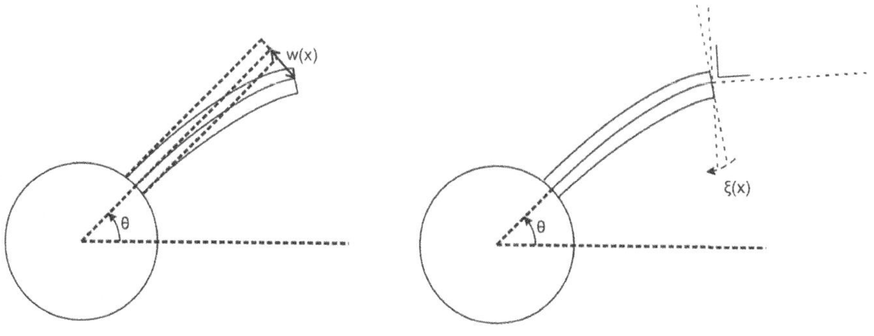

Fig. 2 Deflections of the rotating beam

The beam at $x = 0$ is clamped to motor disk and at $x = 1$ has a free end, from that we have boundary conditions

$$\begin{cases} w(0,t) = \xi(0,t) = 0 \\ w'(1,t) + \xi(1,t) = 0 \\ \xi'(1,t) = 0 \end{cases} \tag{3}$$

for $t > 0$.

After putting that $\rho = \tilde{\rho}A$, $I_\rho = \tilde{\rho}I$ we can convert system (2) into

$$\begin{cases} \ddot{w}(x,t) - \frac{K}{\rho}(w''(x,t) + \xi'(x,t)) = -\ddot{\theta}(t)(r+x), \\ \ddot{\xi}(x,t) - \frac{EA}{\rho}\xi''(x,t) + \frac{K}{I_\rho}(w'(x,t) + \xi(x,t)) + \frac{\tilde{\nu}^2}{I_\rho}\dot{\xi}(x,t) = \ddot{\theta}(t), \end{cases} \tag{4}$$

where I_ρ denotes moment of inertia and ρ—linear density. These parameters are used in mathematical literature. In engineering literature it is assumed furthermore that $K = \kappa GA$, where κ is the shape factor and G is the shear modulus. Then our rotating beam model (4) with engineering parameters takes form

$$\begin{cases} \ddot{w}(x,t) - \frac{\kappa GA}{\rho}(w''(x,t) + \xi'(x,t)) = -\ddot{\theta}(t)(r+x), \\ \ddot{\xi}(x,t) - \frac{EA}{\rho}\xi''(x,t) + \frac{\kappa GA}{I_\rho}(w'(x,t) + \xi(x,t)) + \frac{\tilde{\nu}^2}{I_\rho}\dot{\xi}(x,t) = \ddot{\theta}(t). \end{cases}$$

In further considerations we use the model and parameters defined in (4).

Now we want to normalize the units for simplicity the system (4), i.e., we will denote $\frac{EA}{K} = \gamma^2$ and use an appropriate change of variables (cf. [5])

$$\tilde{t} = \sqrt{\frac{K}{I_\rho}}t, \; \tilde{x} = \sqrt{\frac{\rho}{I_\rho}}x, \tag{5}$$

of constant

$$\tilde{r} = \sqrt{\frac{\rho}{I_\rho}}r, \tag{6}$$

and of functions

$$\tilde{w}(\tilde{x},\tilde{t}) = \sqrt{\frac{\rho}{I_\rho}}w(x,t), \; \tilde{\xi}(\tilde{x},\tilde{t}) = \xi(x,t), \tag{7}$$

$$\tilde{\theta}(\tilde{t}) = \theta(t).$$

Then, system (4) with particular type of viscoelastic damping with respect to ξ can be transferred into

$$\begin{cases} \ddot{\tilde{w}}(\tilde{x},\tilde{t}) - \tilde{w}''(\tilde{x},\tilde{t}) - \tilde{\xi}'(\tilde{x},\tilde{t}) = -\ddot{\tilde{\theta}}(\tilde{t})(\tilde{r}+\tilde{x}), \\ \ddot{\tilde{\xi}}(\tilde{x},\tilde{t}) - \gamma^2\tilde{\xi}''(\tilde{x},\tilde{t}) + \tilde{w}'(\tilde{x},\tilde{t}) + \tilde{\xi}(\tilde{x},\tilde{t}) + \frac{\tilde{\nu}^2}{\sqrt{KI_\rho}}\dot{\tilde{\xi}}(\tilde{x},\tilde{t}) = \ddot{\tilde{\theta}}(\tilde{t}). \end{cases}$$

Finally, after replacing "˜" by " ", defining $u(t) := \ddot{\theta}(t)$ and $\nu^2 := \frac{\tilde{\nu}^2}{\sqrt{KI_\rho}}$ we obtain

$$\begin{cases} \ddot{w}(x,t) - w''(x,t) - \xi'(x,t) = -u(t)(r+x), \\ \ddot{\xi}(x,t) - \gamma^2\xi''(x,t) + w'(x,t) + \xi(x,t) + \nu^2\dot{\xi}(x,t) = u(t). \end{cases} \tag{8}$$

Now following [5], we will consider operator equation of the form

$$\dot{Z} = \begin{pmatrix} \dot{w} \\ \dot{\xi} \\ \ddot{w} \\ \ddot{\xi} \end{pmatrix} = \underbrace{\begin{pmatrix} 0 & I \\ -A & -B_i \end{pmatrix}}_{\mathcal{A}_i} \begin{pmatrix} w \\ \xi \\ \dot{w} \\ \dot{\xi} \end{pmatrix} + \begin{pmatrix} 0 \\ 0 \\ -r-x \\ 1 \end{pmatrix} u(t), \tag{9}$$

$i = 1, 2, 3$ where[1] $I : D\left(A^{\frac{1}{2}}\right) \to H = L^2\left((0,1), \mathbb{R}^2\right)$ is embedding operator, $A : D(A) \to H$ is linear operator defined by

$$A\begin{pmatrix} y \\ z \end{pmatrix} = \begin{pmatrix} -y'' - z' \\ -\gamma^2 z'' + y' + z \end{pmatrix} \tag{10}$$

for $\begin{pmatrix} y \\ z \end{pmatrix} \in D(A)$, where

$$D(A) = \left\{ \begin{pmatrix} y \\ z \end{pmatrix} \in H^2((0,1), \mathbb{R}^2) \,\middle|\, \begin{matrix} y(0) = z(0) = 0 \\ y'(1) + z(1) = z'(1) = 0 \end{matrix} \right\},$$

and $B_i : D(B_i) \to H$ is a symmetric distributed damping operator, $D(\mathcal{A}_i) = D(A) \times D\left(A^{\frac{1}{2}}\right) \subset \mathcal{H} = D\left(A^{\frac{1}{2}}\right) \times H$. In the first case, damping operator B_1 from [20, 22] of the form

$$B_1\begin{pmatrix} y \\ z \end{pmatrix} = \begin{pmatrix} 0 \\ \nu^2 z \end{pmatrix}, \tag{11}$$

[1] $A^{\frac{1}{2}}$ is a "square root" of A, where A is self-adjoint and positive definite operator. $A^{\frac{1}{2}}$ is also self-adjoint and positive definite (see [5]).

where $D(B_1) = D\left(A^{\frac{1}{2}}\right) \supset D(A)$, was considered. In [19], authors studied input–output stability problem with damping operator B_2 of the form

$$B_2\begin{pmatrix} y \\ z \end{pmatrix} = \begin{pmatrix} -\mu^2 y'' \\ 0 \end{pmatrix}, \tag{12}$$

with $D(B_2) = D(A)$. Now, we use it in the context of analysis of the existence of the optimal stability margin. The last considered case concerns the damping operator B_3 from [21], which is an additive combination of B_1 and B_2, that is of the form

$$B_3\begin{pmatrix} y \\ z \end{pmatrix} = \begin{pmatrix} -\mu^2 y'' \\ \nu^2 z \end{pmatrix}, \tag{13}$$

with $D(B_3) = D(B_2) = D(A)$.

3 Spectral Properties of the Operators $\mathcal{A}_i$

We are going to conduct the spectral analysis for operators $\mathcal{A}_i$. In particular we will show that $\mathcal{A}_1$ is a Riesz-spectral operator, i.e., satisfying the following

Definition 3 (See [1]) Suppose that $\mathcal{A}$ is a linear, closed operator on a Hilbert space, $\mathcal{H}$, with simple eigenvalues $\{\lambda_n, n \geqslant 1\}$ and suppose that the corresponding eigenvectors $\{\phi_n, n \geqslant 1\}$ form a Riesz basis in $\mathcal{H}$. If the closure of $\{\lambda_n, n \geqslant 1\}$ is totally disconnected, then we call $\mathcal{A}$ a *Riesz-spectral operator*.

If operator $\mathcal{A}$ is a Riesz-spectral operator it means that spectrum determined growth condition is satisfied (see [1], Theorem 2.3.5). Thanks to it, analysis of optimal stability margins can be based on the location of the spectrum on complex plane. Note that the proper location of spectrum of system operator in general does not guarantee stability [24], nor existence of spectrum part on the axis [16] does not guarantee instability. In some cases it may depend on how fast the eigenvalues approach the imaginary axis (see, e.g., the case of neutral–type systems [17, 18]). Thus one often needs to conduct a more careful analysis.

Detailed proofs of the stated below lemmas and theorems can be found in [19–22]. At the beginning we formulate the following:

Lemma 1 *Operators $\mathcal{A}_i$ are invertible and the inverse operators $\mathcal{A}_i^{-1}$ are compact.*

From Lemma 1, we have

Corollary 1 *Operators $\mathcal{A}_i$ have a compact resolvent and the spectra $\sigma(\mathcal{A}_i)$ are point-wise.*

After introducing scalar product in considered Hilbert space $\mathcal{H}$ of the form (see [5])

$$\langle z_1, z_2\rangle_{\mathcal{H}} = \left\langle A^{\frac{1}{2}} v_1, A^{\frac{1}{2}} v_2\right\rangle_H + \langle w_1, w_2\rangle_H \tag{14}$$

for all $z_1 = (v_1, w_1)^T$ and $z_2 = (v_2, w_2)^T$ in $\mathcal{H}$ with accompanying norm $\|z\|_{\mathcal{H}}^2 = \langle z, z\rangle_{\mathcal{H}}$, we are able to prove that the considered operators are dissipative.

Lemma 2 *Operators $\mathcal{A}_i$ and $-\mathcal{A}_1 - \nu^2 I$ are dissipative.*

Proof We present the proof for the more general situation, i.e., for operator $\mathcal{A}_3$. Using the form of the scalar product (14) and for every $z = (y_1, y_2, y_3, y_4)^T \in \mathcal{H}$, we obtain

$$\langle \mathcal{A}_3 z, z\rangle_{\mathcal{H}} = \underbrace{\left\langle A^{\frac{1}{2}} \begin{pmatrix} y_3 \\ y_4 \end{pmatrix}, A^{\frac{1}{2}} \begin{pmatrix} y_1 \\ y_2 \end{pmatrix}\right\rangle_H - \left\langle A \begin{pmatrix} y_1 \\ y_2 \end{pmatrix}, \begin{pmatrix} y_3 \\ y_4 \end{pmatrix}\right\rangle_H}_{=0}$$

$$- \left\langle B_3 \begin{pmatrix} y_3 \\ y_4 \end{pmatrix}, \begin{pmatrix} y_3 \\ y_4 \end{pmatrix}\right\rangle_H = \mu^2 \int_0^1 y_3''(x) y_3(x) dx - \nu^2 \int_0^1 y_4^2(x) dx$$

$$= -\mu^2 \int_0^1 \left(y_3'(x)\right)^2 dx - \nu^2 \int_0^1 y_4^2(x) dx \le 0.$$

For operators $\mathcal{A}_1$, $\mathcal{A}_2$, and $-\mathcal{A}_1 - \nu^2 I$ the proofs are similar. □

Lemma 1, Corollary 1, and Lemma 2 shows that the operators satisfy the assumptions of Lumer–Phillips theorem.

Theorem 1 *Operators $\mathcal{A}_i$ and $-\mathcal{A}_1 - \nu^2 I$ generate contraction semigroups. Hence, $\mathcal{A}_1$ is the infinitesimal generator of a C_0-group.*

Existence of solution and well–posedness of systems of type (9) follow from Theorem 1.

Theorem 2 *C_0-semigroups $T_i(t)$ generated by $\mathcal{A}_i$ are asymptotically stable.*

The proof of Theorem 2 follows from Lyubich and Phóng's theorem [9, 16]. It is sufficient to show that $\mathrm{Re}\lambda < 0$ for any $\lambda \in \sigma(\mathcal{A}_i)$. The dissipativity of $\mathcal{A}_i$ (Lemma 2) implies $\mathrm{Re}\lambda \le 0$ for any $\lambda \in \sigma(\mathcal{A}_i)$. Hence it suffices to see that there are no eigenvalues on the imaginary axis, that is an easy exercise.

Now we find general form of spectral equations. We introduce lemma, which helps us in further considerations, for two main cases of physical parameters of the beam, $\gamma^2 > 1$ and $\gamma^2 = 1$, and different values of damping coefficients μ and ν.

Lemma 3 (See [19–22]) *Spectral equation $\mathcal{P}_i(\lambda) = 0$ of system* (9) *can be written in the form*

$$\mathcal{P}_i(\lambda) = (a_{23}^{(i)}(1, \lambda) + a_{33}^{(i)}(1, \lambda)) a_{44}^{(i)}(1, \lambda) - (a_{24}^{(i)}(1, \lambda) + a_{34}^{(i)}(1, \lambda)) a_{43}^{(i)}(1, \lambda) = 0,$$

where $a^{(i)}_{jk}(1,\lambda)$ *are in each case elements of matrix exponential of*

$$M_1(\lambda) = \begin{pmatrix} 0 & 0 & 1 & 0 \\ 0 & 0 & 0 & 1 \\ \lambda^2 & 0 & 0 & -1 \\ 0 & \frac{\lambda^2+\nu^2\lambda+1}{\gamma^2} & \frac{1}{\gamma^2} & 0 \end{pmatrix}, \quad M_2(\lambda) = \begin{pmatrix} 0 & 0 & 1 & 0 \\ 0 & 0 & 0 & 1 \\ \frac{\lambda^2}{1+\mu^2\lambda} & 0 & 0 & -\frac{1}{1+\mu^2\lambda} \\ 0 & \lambda^2+1 & 1 & 0 \end{pmatrix},$$

$$M_3(\lambda) = \begin{pmatrix} 0 & 0 & 1 & 0 \\ 0 & 0 & 0 & 1 \\ \frac{\lambda^2}{1+\mu^2\lambda} & 0 & 0 & -\frac{1}{1+\mu^2\lambda} \\ 0 & \lambda^2+\nu^2\lambda+1 & 1 & 0 \end{pmatrix}.$$

4 Asymptotic Spectral Analysis of Operators $\mathcal{A}_i$

Now we will go to find asymptotic formulas for eigenvalues of operators $\mathcal{A}_i$, $i = 1, 2, 3$ with different type of damping operators B_1, B_2, and B_3. In particular, one can easily see that each $\mathcal{A}_i$ has only a point spectrum of finite multiplicity. Then, we will observe that C_0–group $T_1(t)$ generated by $\mathcal{A}_1$ satisfies the spectrum determined growth condition.

4.1 Operator $\mathcal{A}_1$

4.1.1 Case $\mu = 0$, $\nu > 0$, $\gamma^2 > 1$

At the beginning we analyze the case of $\gamma^2 > 1$ and arbitrary value of damping coefficient ν.

Let $\nu > 0$ and $\gamma^2 > 1$, then the considered model has the following form:

$$\begin{cases} \ddot{w}(x,t) - w''(x,t) - \xi'(x,t) = -u(t)(r+x), \\ \ddot{\xi}(x,t) - \gamma^2\xi''(x,t) + w'(x,t) + \xi(x,t) + \nu^2\dot{\xi}(x,t) = u(t). \end{cases} \tag{15}$$

for $x \in (0,1)$ and $t > 0$ with boundary conditions (3).

Theorem 3 *Let* $\gamma^2 > 1$. *For any value of a damping constant* $0 < \nu < \infty$ *the eigenvalues of the operator* $\mathcal{A}_1$ *are two asymptotic families* $\lambda^{(1)}_k = -\frac{1}{2}\nu^2 + \gamma\frac{2k+1}{2}\pi i + \varepsilon^{(1)}_k$ *and* $\lambda^{(2)}_k = \frac{2k+1}{2}\pi i + \varepsilon^{(2)}_k$, *where* $\lim_{k\to\infty} \varepsilon^{(i)}_k = 0$ *(see Figs. 3, 4, 5, and 6).*

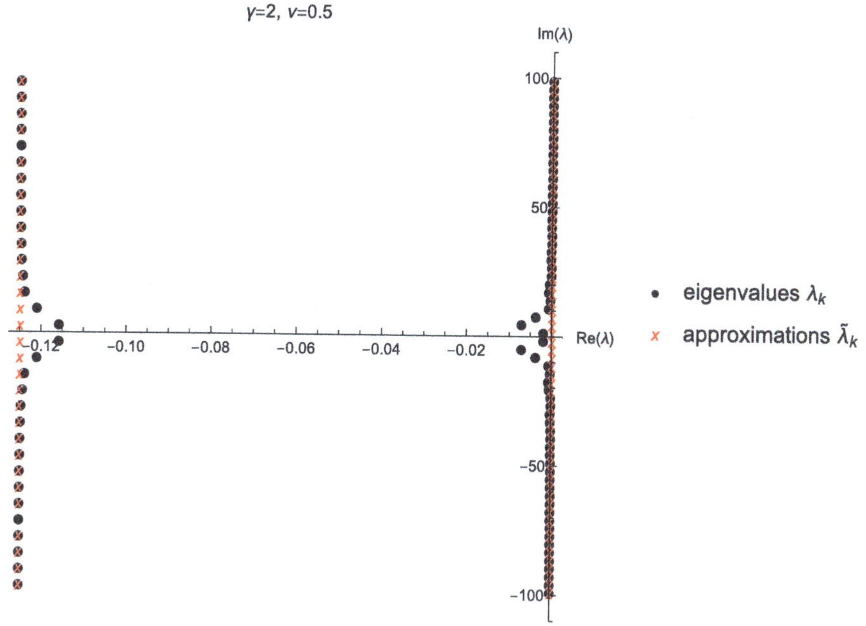

Fig. 3 Eigenvalues in the case $\nu = 0.5$ and $\gamma = 2$

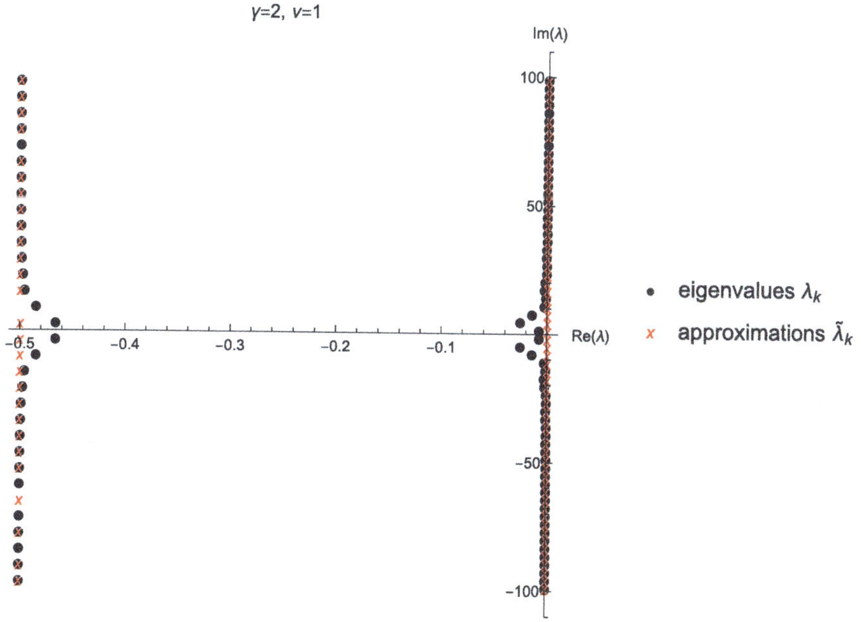

Fig. 4 Eigenvalues in the case $\nu = 1$ and $\gamma = 2$

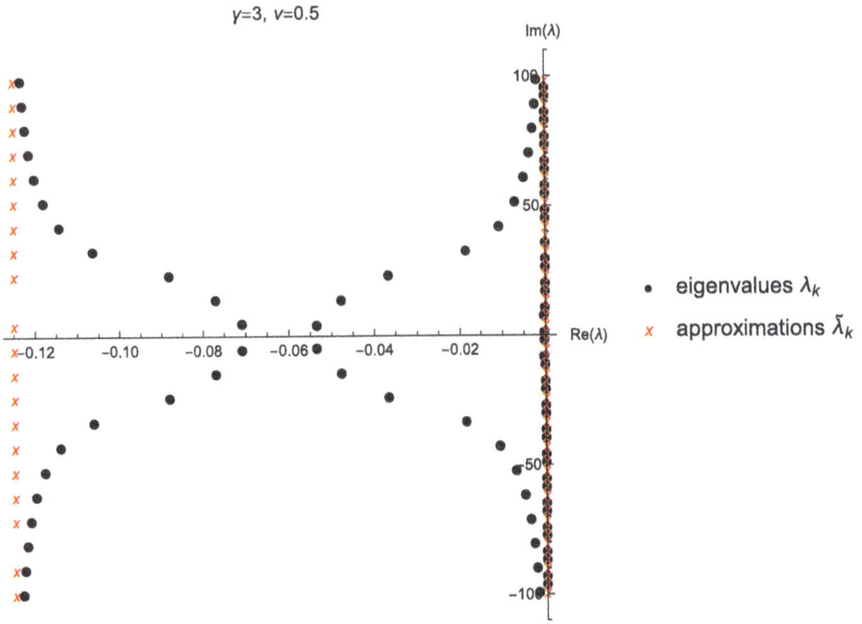

Fig. 5 Eigenvalues in the cases $\nu = 0.5$ and $\gamma = 3$

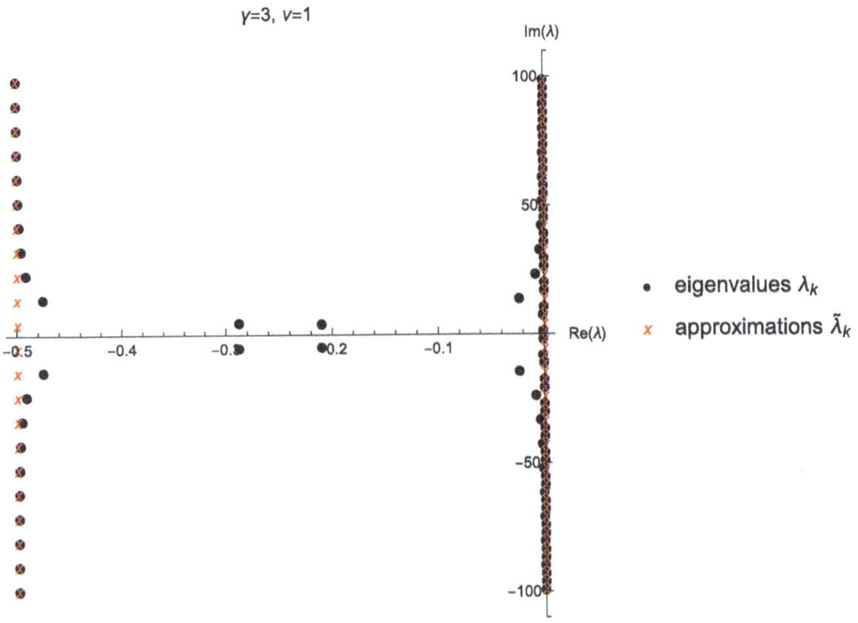

Fig. 6 Eigenvalues in the cases $\nu = 1$ and $\gamma = 3$

4.1.2 Case $\mu = 0, \nu > 0, \gamma^2 = 1$

Now, we examine the particular model of a rotating Timoshenko beam, where physical constant $\gamma^2 = \frac{EA}{K} = 1$. For any $\nu > 0$, we obtain

$$\begin{cases} \ddot{w}(x,t) - w''(x,t) - \xi'(x,t) = -u(t)(r+x), \\ \ddot{\xi}(x,t) - \xi''(x,t) + w'(x,t) + \xi(x,t) + \nu^2\dot{\xi}(x,t) = u(t). \end{cases} \tag{16}$$

for $x \in (0,1)$ and $t > 0$ with boundary conditions (3).

Theorem 4 *Let $\gamma^2 = 1$. For any value of a damping constant $0 < \nu < \infty$ the system* (16) *has the spectrum of the following form:*

$$\begin{cases} \lambda_k^{(1)} = -\frac{1}{4}\nu^2 + \frac{1}{2}\ln\left(y - \sqrt{y^2-1}\right) + \frac{2k+1}{2}\pi i + \varepsilon_k^{(1)}, \\ \lambda_k^{(2)} = -\frac{1}{4}\nu^2 + \frac{1}{2}\ln\left(y + \sqrt{y^2-1}\right) + \frac{2k+1}{2}\pi i + \varepsilon_k^{(2)}, \end{cases}$$

where $\lim_{k\to\infty} \varepsilon_k^{(i)} = 0$ (see Figs. 7 and 8), and $y = y(\nu)$ is given by

$$y = \frac{\nu^4 \cosh\left(\frac{\sqrt{\nu^4-4}}{2}\right) - 4}{\nu^4 - 4}. \tag{17}$$

4.2 Operator $\mathcal{A}_2$

Analysis of spectrum of operator $\mathcal{A}_1$ in the case described in Sect. 4.1.1 shows that for physical parameter $\gamma^2 > 1$ one family of eigenvalues is asymptotically close to imaginary axis, thus no stability margin may be expected. Therefore, in the further analysis we will omit cases where $\gamma^2 > 1$ and we will analyze only those with $\gamma^2 = 1$.

4.2.1 Case $\mu > 0, \nu = 0, \gamma^2 = 1$

Let $\mu > 0$, system (9) with damping operator (12) can be written as

$$\begin{cases} \ddot{w}(x,t) - \mu^2\dot{w}''(x,t) - w''(x,t) - \xi'(x,t) = -u(t)(r+x) \\ \ddot{\xi}(x,t) - \xi''(x,t) + w'(x,t) + \xi(x,t) = u(t) \end{cases} \tag{18}$$

for $x \in (0,1)$ and $t > 0$ with boundary conditions (3).

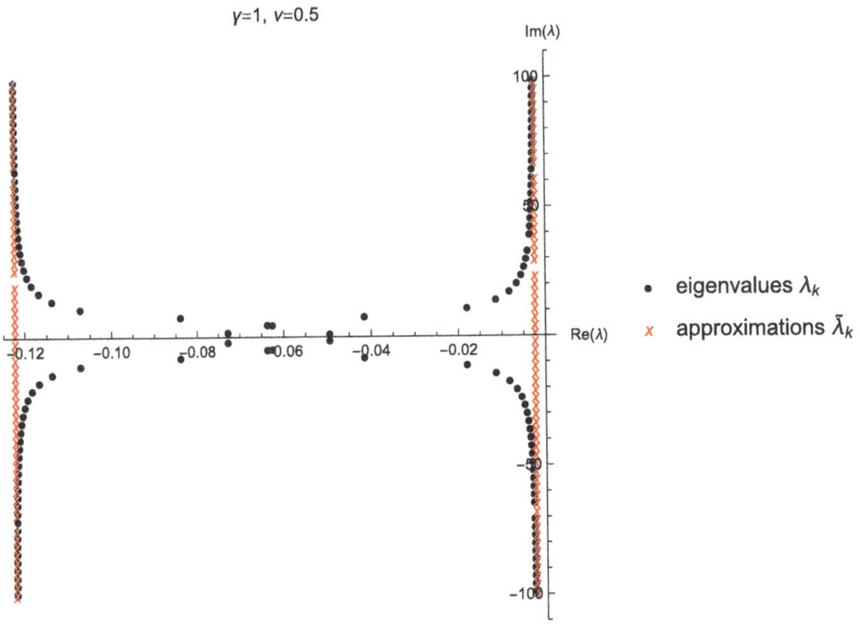

Fig. 7 Eigenvalues in the case $\nu = 0.5$ and $\gamma = 1$

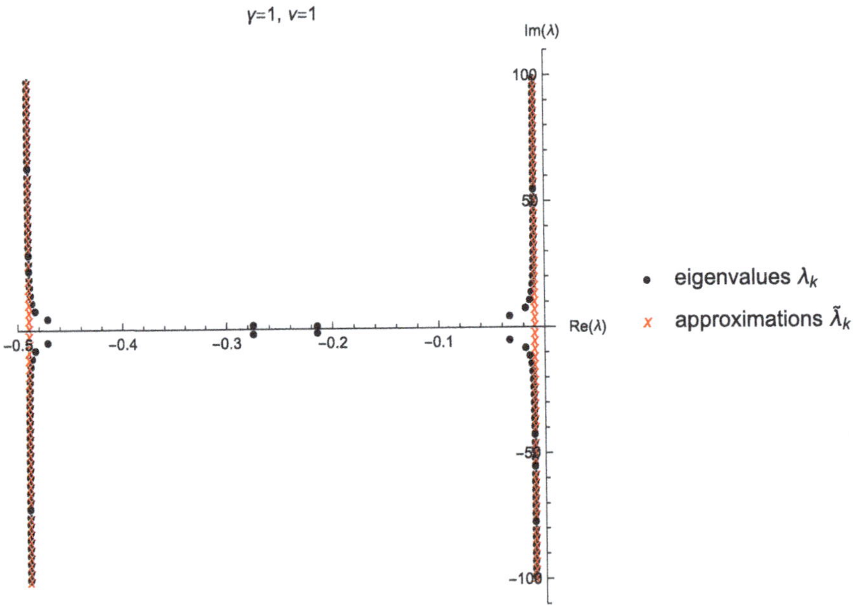

Fig. 8 Eigenvalues in the case $\nu = 1$ and $\gamma = 1$

Theorem 5 *Let $\gamma^2 = 1$. For any value of a damping constant $0 < \mu < \infty$ the eigenvalues of operator $\mathcal{A}_2$ form two asymptotic families $\lambda_k^{(1)} = -\nu^2\left(\frac{2k+1}{2}\pi\right)^2 + \varepsilon_k^{(1)}$ and $\lambda_k^{(2)} = \frac{2k+1}{2}\pi i + \varepsilon_k^{(2)}$, where $\lim_{k\to\infty} \varepsilon_k^{(i)} = 0$ (see Figs. 9 and 10).*

4.3 Operator $\mathcal{A}_3$

Similarly as in previous section, we should consider only the case with physical parameter $\gamma^2 = 1$.

4.3.1 Case $\mu > 0, \nu > 0, \gamma^2 = 1$

To complete our consideration with $\gamma^2 = 1$, we examine the case with additive combination of damping operators (11) and (12), i.e., damping operator (13).

Theorem 6 *Let $\gamma^2 = 1$. For any value of a damping constant $0 < \mu < \infty$ the eigenvalues of operator $\mathcal{A}_3$ consist of two asymptotic families $\lambda_k^{(1)} =$*

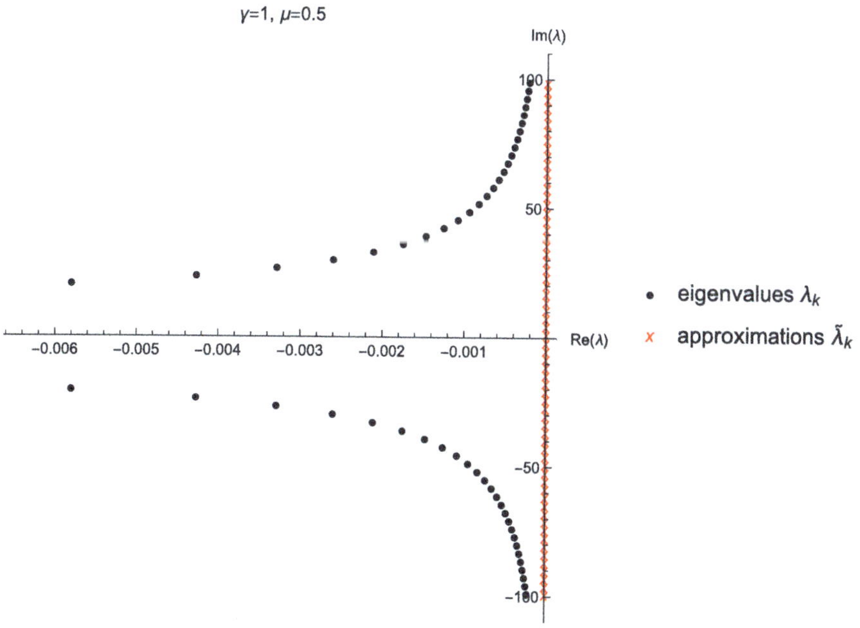

Fig. 9 Eigenvalues in the case $\mu = 0.5$ and $\gamma = 1$ (only family $\lambda_k^{(2)}$ shown)

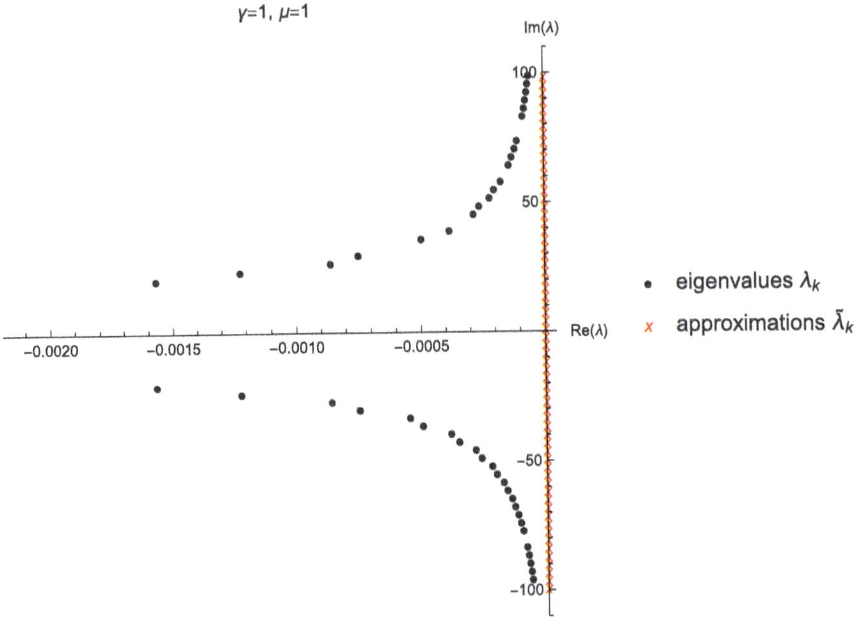

Fig. 10 Eigenvalues in the case $\mu = 1$ and $\gamma = 1$ (only family $\lambda_k^{(2)}$ shown)

$-\mu^2\left(\frac{2k+1}{2}\pi\right)^2 + \varepsilon_k^{(1)}$ *and* $\lambda_k^{(2)} = -\frac{1}{2}\nu^2 + \frac{2k+1}{2}\pi i + \varepsilon_k^{(2)}$, *where* $\lim_{k\to\infty} \varepsilon_k^{(i)} = 0$ *(see Figs. 11 and 12).*

Analysis of all the considered cases allows us to formulate the following corollaries.

Corollary 2 (See [22]) *Each eigensystem of operators* $\mathcal{A}_i$ *form a complete set in* $\mathcal{H}$.

Corollary 3 *For any value of a damping constant* $0 < \mu, \nu < \infty$ *spectra* $\sigma(\mathcal{A}_i)$ *consist of isolated single eigenvalues of* $\mathcal{A}_i$.

The last requirement for the considered operator $\mathcal{A}_1$ to be a Riesz-spectral operator is that the eigensystem of $\mathcal{A}_1$ forms a Riesz basis on $\mathcal{H}$. With the help comes the following theorem

Theorem 7 (See [23, 25]) *Let* $\mathcal{A}$ *be the infinitesimal generator of the* C_0*-group* $(T(t))_{t\in\mathbb{R}}$ *on the Hilbert space* $\mathcal{H}$. *We denote the eigenvalues of* $\mathcal{A}$ *by* λ_n *(counting with multiplicity), and the corresponding (normalized) eigenvectors by* $\{\phi_n\}$. *If the following two conditions hold,*

1. *The span of the eigenvectors forms a dense set in* $\mathcal{H}$.

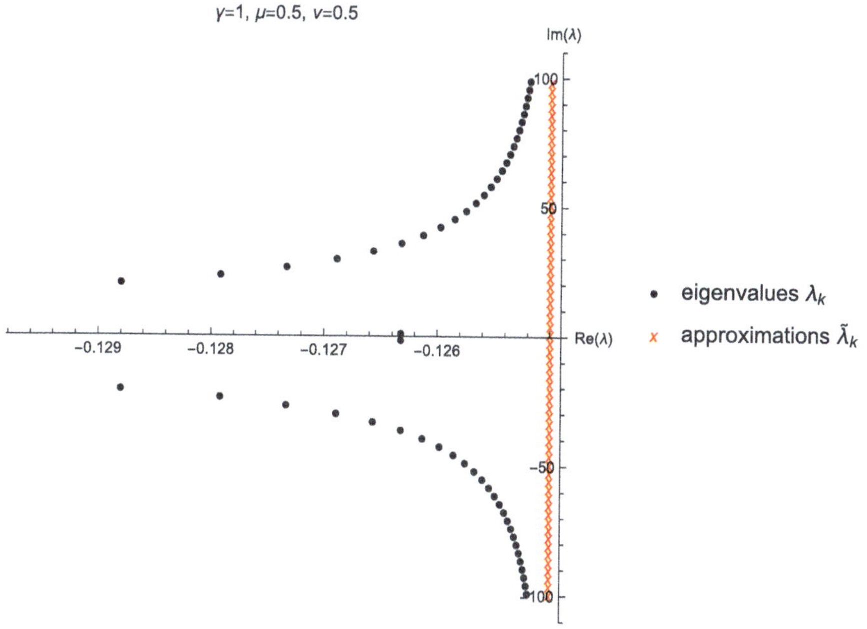

Fig. 11 Eigenvalues in the case $\mu = 0.5$, $\nu = 0.5$, and $\gamma = 1$ (only family $\lambda_k^{(2)}$ shown)

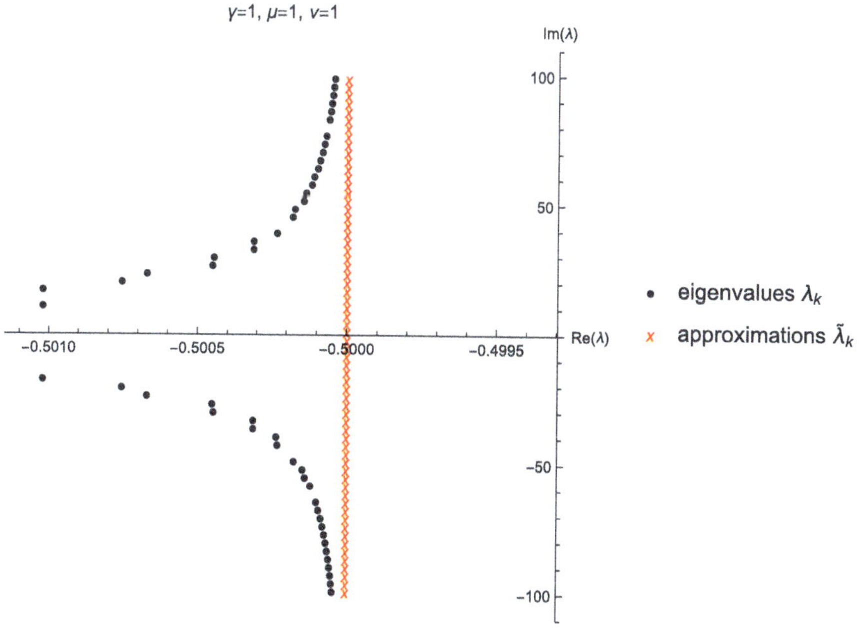

Fig. 12 Eigenvalues in the case $\mu = 1$, $\nu = 1$, and $\gamma = 1$ (only family $\lambda_k^{(2)}$ shown)

2. The point spectrum has a uniform gap, *i.e.,*

$$\inf_{n \neq m} |\lambda_n - \lambda_m| > 0,$$

then the eigenvectors form a Riesz basis on $\mathcal{H}$.

Corollary 4 *Taking into account information from Sect. 3, Corollary 3, and the above theorem we can observe that $\mathcal{A}_1$ is a Riesz-spectral operator, which means that the spectrum determined growth condition is satisfied.*

5 Stability Margin Analysis

Now we will analyze stability margins of the cases considered in Sect. 4. In the case described in Sects. 4.1.1 and 4.2.1 adding damping operator to a Timoshenko beam model caused partial exponential stability. It means that there exist states (subspaces) of the system for which the norm vanishes exponentially, and there exist states for which it is asymptotically but not exponentially stable. From Theorem 2 we know that the imaginary axis is the asymptote of the spectrum of $\mathcal{A}_1$ (for $\gamma^2 > 1$) and $\mathcal{A}_2$, this implies that the system is not exponentially stable. In these cases stability margin is equal to 0 (see Fig. 13 for the case of $\mathcal{A}_1$ with $\gamma^2 > 1$).

In the case described in Sect. 4.3.1 we can see that $\mathrm{Re}(-\lambda) \geq \min \left\{\mu^2 \left(\frac{2k+1}{2}\pi\right)^2, \frac{1}{2}\nu^2\right\}$. There is no reason of looking for optimal stability margin here because stability margin of the system goes to infinity as damping coefficients go to infinity.

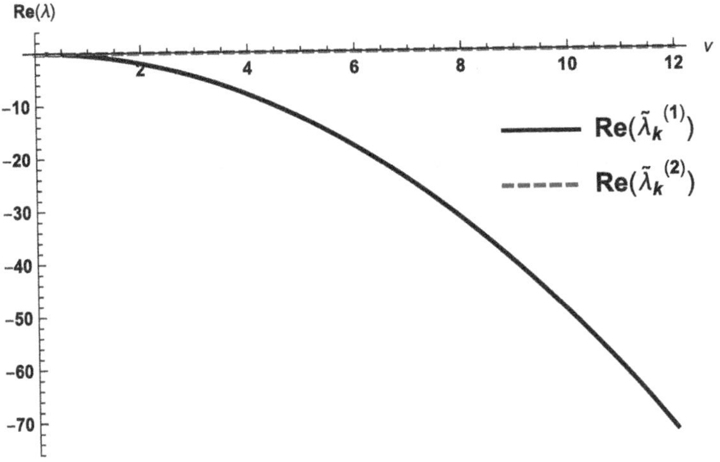

Fig. 13 Plots of $\mathrm{Re}\left(\tilde{\lambda}_k^{(1)}\right)$ (solid line) and $\mathrm{Re}\left(\tilde{\lambda}_k^{(2)}\right)$ (dashed line) for $\mu = 0$, $\nu > 0$, $\gamma^2 > 1$

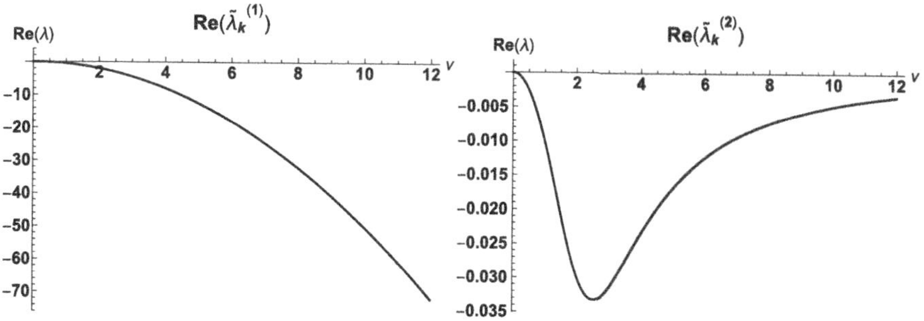

Fig. 14 Plots of Re $\left(\tilde{\lambda}_k^{(1)}\right)$ (left) and Re $\left(\tilde{\lambda}_k^{(2)}\right)$ (right) in the case $\gamma^2 = 1$; note the different scales on the plots

Now we proceed to the case from Sect. 4.1.2. At the beginning we analyze the behavior of eigenvalues with increase of the damping coefficient ν (see Fig. 14). We can observe that increasing of a damping coefficient ν caused that the first family is moving away from imaginary axis (see Fig. 14 for Re $\left(\tilde{\lambda}_k^{(1)}\right)$), while for the second family there exists the optimal damping coefficient. Increasing ν above the optimal value causes slower energy dissipation (overdamping). Comparing those results according to Definition 2, we can plot stability margin dependence on the damping coefficient ν.

Corollary 5 *The optimal stability margin of system* (9) *with operator* $\mathcal{A}_1$ *(under assumption* $\gamma^2 = 1$*) is* $\omega_0 = -0.03324163912497735136$ *(for* $\nu_{\text{opt}} = 2.54189087636624306026$*) (see Fig. 15).*

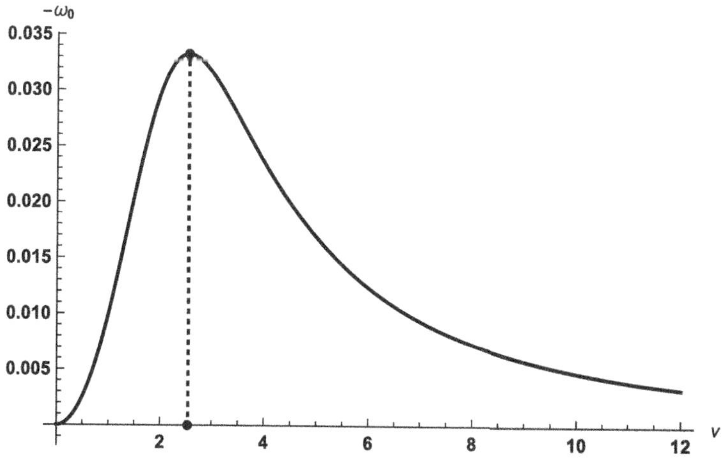

Fig. 15 Optimal stability margin (case $\mu = 0$, $\nu > 0$, $\gamma^2 = 1$)

6 Conclusions

We analyzed stability of a slowly vibrating Timoshenko beam with a distributed damping connected to rotations of cross-sections of the beam, of deflections of the center line of the beam, and of both.

For the system with operator $\mathcal{A}_2$ there appears no stability margin for any value of a damping parameter μ.

For the system with operator $\mathcal{A}_3$ there exists a stability margin which is enlarging with increase of any of damping parameters μ, ν, thus there is no reason of looking for optimal stability margin.

For the system with operator $\mathcal{A}_1$ we observe two different possible behaviors—in the case of $\gamma^2 > 1$ there is no stability margin again, and in the case of $\gamma^2 = 1$ there appear underdamping, overdamping, and critical damping effects, and we were able to find the value of optimal damping parameter ν_0.

In the above cases where there was no exponential stability margin at all it is still an unresolved open question of possible polynomial stability phenomenon (cf. [11, 14]).

References

1. Curtain, R.F., Zwart, H.J.: An Introduction to Infinite-Dimensional Linear Systems Theory. Springer, Berlin (1995)
2. Cveticanin, L.: Oscillators with nonlinear elastic and damping forces. Comput. Math. Appl. **62**(4), 1745–1757 (2011)
3. Iannelli, M.: Mathematical problems in the description of age structured populations. In: Capasso, V., Grosso, E., Paveri-Fontana, S.L. (eds.) Mathematics in Biology and Medicine, pp. 19–32. Springer, Berlin (1985)
4. Ji, C., Jiang, D., Shi, N.: The behavior of an SIR epidemic model with stochastic perturbation. Stoch. Anal. Appl. **30**(5), 755–773 (2012)
5. Krabs, W., Sklyar, G.M.: On Controllability of Linear Vibrations. Nova Science Publishers, Huntington (2002)
6. Levin, S.A.: Population Models and Community Structure in Heterogeneous Environments, pp. 295–320. Springer, Berlin (1986)
7. Liu, Q., Ruan, D., Huang, X.: Topology optimization of viscoelastic materials on damping and frequency of macrostructures. Comput. Method. Appl. Mech. Eng. **337**, 305–323 (2018)
8. Liu, J., Guan, J., Wan, X., Shang, R., Shi, X., Fang, L., Liu, C.: The improved cargo loading and physical stability of ibuprofen orodispersible film: molecular mechanism of ion-pair complexes on drug-polymer miscibility. J. Pharm. Sci. **109**(3), 1356–1364 (2020)
9. Lyubich, Y.I., Phóng, V.Q.: Asymptotic stability of linear differential equations in Banach spaces. Stud. Math. **88**, 34–37 (1988)
10. Ma, F., Imam, A., Morzfeld, M.: The decoupling of damped linear systems in oscillatory free vibration. J. Sound Vib. **324**(1), 408–428 (2009)
11. Mercier, D., Régnier, V.: Decay rate of the Timoshenko system with one boundary damping. Evol. Equ. Control Theory **8**(2), 423–445 (2019)
12. Migdalovici, M., Vlădăreanu, L., Baran, D., Vlădeanu, G., Radulescu, M.: Stability analysis of the walking robots motion. Proc. Comput. Sci. **65**, 233–240 (2015)

13. Rohrig, T.P.: Drug stability. In: T.P. Rohrig (ed.) Postmortem Toxicology, pp. 101–121. Academic Press, Cambridge (2019)
14. Santos, M.L., Júnior, D.S.A., Rivera, J.E.M.: The stability number of the Timoshenko system with second sound. J. Differ. Equ. **253**(9), 2715–2733 (2012)
15. Shao, C., Togashi, J., Mitobe, K., Capi, G.: Utilizing the nonlinearity of tendon elasticity for compensation of unknown gravity of payload. J. Robot. Mechatron. **30**(6), 873–879 (2018)
16. Sklyar, G., Shirman, V.Y.: On the asymptotic stability of a linear differential equation in a Banach space. Teor. Funkc. Anal. Prilozh. **37**, 127–132 (1982)
17. Sklyar, G.M., Polak, P.: Asymptotic growth of solutions of neutral type systems. Appl. Math. Optim. **67**, 453–477 (2013)
18. Sklyar, G.M., Polak, P.: On asymptotic estimation of a discrete type C_0-semigroups on dense sets: application to neutral type systems. Appl. Math. Optim. **75**(2), 175–192 (2017)
19. Woźniak, J., Firkowski, M.: Note on the stability of a slowly rotating Timoshenko beam with damping. Adv. Appl. Math. Mech. **7**(6), 736–753 (2015)
20. Woźniak, J., Firkowski, M.: Optimal damping coefficient of a slowly rotating Timoshenko beam. In: Proceedings of the SIAM Conference on Control and its Applications, pp. 81–84 (2015)
21. Woźniak, J., Firkowski, M.: Stability of slowly rotating Timoshenko beam with two viscoelastic damping coefficients. In: Proceedings of the 23rd MED Conference Control and Automation, pp. 404–407 (2015)
22. Woźniak, J., Firkowski, M.: Optimal decay ratio of damped slowly rotating Timoshenko beams. Z. Angew. Math. Mech. **99**(10), e201800222 (2019)
23. Xu, G.Q., Yung, S.P.: The expansion of a semigroup and a Riesz basis criterion. J. Differ. Equ. **210**, 1–24 (2005)
24. Zabczyk, J.: A note on C_0-semigroups. Bull. Pol. Acad. Sci. **23**, 895–898 (1975)
25. Zwart, H.: Riesz basis for strongly continuous groups. J. Differ. Equ. **249**, 2397–2408 (2010)

Stabilization of Crystallization Models Governed by Hyperbolic Systems

Alexander Zuyev and Peter Benner

Abstract This chapter deals with mathematical models of continuous crystallization described by hyperbolic systems of partial differential equations coupled with ordinary and integro-differential equations. The considered systems admit nonzero steady-state solutions with constant inputs. To stabilize these solutions, we present an approach for constructing control Lyapunov functionals based on quadratic forms in weighted L^2-spaces. It is shown that the proposed control design scheme guarantees exponential stability of the closed-loop system.

1 Introduction

The study of the literature in the field of mathematical control theory for distributed parameter systems shows that the development of control design techniques is to a considerable extent influenced by problems of chemical engineering. Important examples in this area come from mathematical models of distillation, chromatography, and crystallization processes governed by hyperbolic systems of partial differential equations [1–4]. For a moving bed chromatography with considerable apparent dispersion coefficients, a parabolic type equilibrium dispersive model is also available for theoretical studies (cf. [5]).

A. Zuyev (✉)
Max Planck Institute for Dynamics of Complex Technical Systems, Magdeburg, Germany

Institute of Applied Mathematics and Mechanics, National Academy of Sciences of Ukraine, Sloviansk, Ukraine

Donbas State Pedagogical University, Sloviansk, Ukraine
e-mail: zuyev@mpi-magdeburg.mpg.de

P. Benner
Max Planck Institute for Dynamics of Complex Technical Systems, Magdeburg, Germany
e-mail: benner@mpi-magdeburg.mpg.de

G. Sklyar, A. Zuyev (eds.), *Stabilization of Distributed Parameter Systems: Design Methods and Applications*, SEMA SIMAI Springer Series 2,
https://doi.org/10.1007/978-3-030-61742-4_8

The main challenge concerning applications of Lyapunov's direct method to quasilinear hyperbolic systems is related to the construction of a Lyapunov functional with negative definite time derivative. For a class of hyperbolic systems with boundary control, strict control Lyapunov functionals have been proposed in [6]. The construction of these functionals requires that the solution of an associated ordinary differential equation should be defined on a prescribed interval. The proposed method has been applied, in particular, to stabilize the equilibrium of the Saint-Venant equations represented as a 2×2 hyperbolic system. An important feature of this approach relies on the possibility of studying control systems with nonuniform steady states.

It should be mentioned that the backstepping approach [7] has already been applied for solving the stabilization problem for several classes of distributed parameter systems [8]. In particular, this approach has been developed in [9] for problems of trajectory generation and tracking for linear 2×2 hyperbolic systems of partial differential equations with boundary inputs and outputs. In the paper [10], the backstepping approach is applied to the output regulation problem for a class of coupled linear parabolic integro-differential equations. To the best of our knowledge, this design methodology has not been applied for the exponential stabilization of integro-differential models of cooling and preferential crystallization so far.

The flatness-based approach [11] is shown to be a powerful method for nonlinear models of chemical engineering with known flat output (see, e.g., [12]). This approach is also applicable for the trajectory tracking problem of distributed parameter systems with integral terms, including a class of parabolic-like linear Volterra partial integro-differential equation with boundary control [13]. However, the question of checking flatness for general classes of systems and constructing a flat output remains open up to now.

Although there are well-established control design techniques for hyperbolic systems with boundary controls [6], mathematical models of crystallization processes require the analysis of coupled systems of first-order quasilinear partial and ordinary differential equations with integral terms. A solution of the local steering problem for a finite-dimensional nonlinear crystallization model has been proposed in [14] by exploiting the Lie bracket approximation techniques with open-loop controls (cf. [15]). For infinite-dimensional crystallization models, the design of stabilizing feedback laws remains an open problem. This chapter aims at solving this problem for the classes of continuous crystallization models introduced in [16] and [17].

2 Continuous Crystallization Model

Consider a continuous cooling crystallization model described by the population balance and mass balance equations as follows [16]:

$$\begin{aligned}&\frac{\partial n(x,t)}{\partial t}+G(x,c)\frac{\partial n(x,t)}{\partial x}=v\psi(x)n(x,t), \quad x\in[0,\ell],\\&n(0,t)=B(c)/G(0,c), \quad G>0,\ B\geq 0,\end{aligned} \tag{1}$$

$$\begin{aligned}\frac{dc}{dt}&=(\rho_0-c)\left(v+\frac{d\ln\varepsilon(n(\cdot,t))}{dt}\right)\\&+\frac{v}{\varepsilon(n(\cdot,t))}\left(u_f-\rho_0-\rho_0 k_v\int_0^\ell \phi(x)n(x,t)dx\right),\end{aligned} \tag{2}$$

$$\varepsilon(n(\cdot,t))=1-k_v\int_0^\ell x^3 n(x,t)dx>0.$$

Here, the crystal size distribution function $n(x,t)\in\mathbb{R}^+=[0,+\infty)$ denotes the expected number of crystals of size $x\in[0,\ell]$ at time $t\geq 0$. Equation (2) relates the solid phase with the mass concentration of solute $c=c(t)\geq 0$ in the liquid phase, where $\rho_0>0$ is the crystal density, $v>0$ is the flow-rate parameter, and $\varepsilon(n(\cdot,t))$ is the void fraction. The crystallization process is controlled by the mass concentration of the solute in the feed $u_f\geq 0$. In this chapter, we allow the growth rate $G(x,c)$ to depend on the crystal size. We refer the reader to [16] for information about the nucleation rate $B(c)$, classification functions $\psi(x)$, $\phi(x)$, and the volumetric shape factor k_v. The functions $G(x,c)$ and $B(c)$ are assumed to be continuously differentiable in their domains of definition, while $\psi(x)$ and $\phi(x)$ are piecewise continuous.

Equations (1) and (2) admit the steady-state solution $n(x,t)=\bar{n}(x)$ and $c(t)=\bar{c}$ with a constant control $u_f=\bar{u}_f$, where

$$\begin{aligned}\bar{n}(x)&=\frac{B(\bar{c})}{G(0,\bar{c})}\exp\left\{v\int_0^x\frac{\psi(y)dy}{G(y,\bar{c})}\right\}, \quad x\in[0,\ell],\\\bar{c}&=\rho_0+\frac{1}{\varepsilon(\bar{n}(\cdot))}\left(\bar{u}_f-\rho_0-\rho_0 k_v\int_0^\ell\phi(x)\bar{n}(x)dx\right).\end{aligned} \tag{3}$$

Our goal is to stabilize the above equilibrium by a state feedback law. By performing the change of variables

$$\begin{aligned}n(x,t)&=\bar{n}(x)+w(x,t),\\c(t)&=\bar{c}+s(t),\\u_f&=\bar{u}_f+u,\end{aligned}$$

we represent the linear approximation of (1) and (2) in a neighborhood of (3) as follows:

$$
\begin{aligned}
&\frac{\partial w(x,t)}{\partial t} = -g(x)w'(x,t) + v\psi(x)w(x,t) - g_c(x)\bar{n}'(x)s(t), \quad x \in [0,\ell],\\
&w(0,t) = \alpha s(t),\\
&\frac{ds(t)}{dt} = -k_0 s(t) + k_1 w(\ell,t) + \int_0^\ell \theta(x)w(x,t)dx + bu,
\end{aligned} \tag{4}
$$

where the prime stands for the derivative with respect to x,

$$
\begin{aligned}
&g(x) = G(x,\bar{c}),\ g_c(x) = \left.\frac{\partial G(x,c)}{\partial c}\right|_{c=\bar{c}},\ \alpha = \frac{d}{dc}\left.\left(\frac{B(c)}{G(0,c)}\right)\right|_{c=\bar{c}},\\
&k_0 = v + \frac{(\bar{c}-\rho_0)k_v}{\varepsilon(\bar{n})}\int_0^\ell x^3 g_c \bar{n}' dx,\ k_1 = \frac{(\rho_0-\bar{c})k_v\ell^3}{\varepsilon(\bar{n})}g(\ell),\\
&\theta(x) = \frac{k_v}{\varepsilon(\bar{n})}\left\{(\bar{c}-\rho_0)\big((x^3 g(x))' + vx^3\psi(x)\big) - v\rho_0\phi(x) + v\beta x^3\right\},\\
&\beta = \bar{u}_f - \rho_0 - \rho_0 k_v \int_0^\ell \phi(x)\bar{n}(x)dx,\ b = \frac{v}{\varepsilon(\bar{n})}.
\end{aligned} \tag{5}
$$

Note that the coefficients and parameters of (4) satisfy the following inequalities for the realistic crystallization example considered in [16]:

$$\rho_0 > \bar{c} > 0,\ k_0 > 0,\ k_1 > 0,\ \alpha > 0,\ b > 0,\ g > 0,\ g_c > 0,\ \psi \le 0.$$

Moreover, the growth rate G is independent of x and affine in c for the example of [16].

3 Control Design

Consider a control Lyapunov functional candidate

$$V = \frac{1}{2}\int_0^\ell \rho(x)w^2(x,t)dx + \frac{\gamma}{2}s^2(t), \tag{6}$$

where $\rho(x) > 0$ is a continuous density function to be defined later, and γ is a positive constant. The time derivative of V along the classical solutions of (4) takes the form

$$\dot{V} = \frac{1}{2}\int_0^\ell \left\{(\rho g)' + 2v\rho\psi\right\} w^2 dx - \left.\frac{\rho g w^2}{2}\right|_{x=\ell} - \left(\gamma k_0 - \frac{\rho(0)g(0)\alpha^2}{2}\right)s^2 + \gamma bsu$$
$$+ s\left(\int_0^\ell (\gamma\theta - \rho g_c \bar{n}')w\,dx + \gamma k_1 w|_{x=\ell}\right). \tag{7}$$

The above formula is obtained by performing the integration by parts with regard to the boundary condition $w(0, t) = \alpha s$. Note that by constructing the Lyapunov functional (6), we aim to achieve strong stability in the corresponding weighted L^2-space. A weaker stability notion with respect to some integral measure has been analyzed in the paper [18] for a population balance model, which is relevant to the stability problem with respect to two measures (cf. [19]) or partial stability concept [20, 21].

It will be shown in the sequel that $\dot{V}$ can be made negative definite in an appropriate state space with the following feedback law:

$$u = -\frac{1}{\gamma b}\left(\varkappa s + \int_0^\ell (\gamma\theta - \rho g_c \bar{n}')w\,dx + \gamma k_1 w|_{x=\ell}\right), \tag{8}$$

where $\varkappa \in \mathbb{R}$ is a design parameter. To answer the question whether the proposed feedback control (8) stabilizes the trivial solution of (4), we take the density function $\rho(x) > 0$ as a solution of the ordinary differential equation

$$\frac{d}{dx}(\rho(x)g(x)) + 2v\psi(x)\rho(x) = -h(x)\rho(x) \quad x \in [0, \ell], \tag{9}$$

with some continuous function $h(x) > 0$ to be defined on $[0, \ell]$. The above equation is a particular case of the differential inequality proposed in [22].

Straightforward computations show that the general solution of (9) is

$$\rho(x) = \bar{\rho}\exp\left\{-\int_0^x \frac{2v\psi(y) + g'(y) + h(y)}{g(y)}dy\right\}, \quad \bar{\rho} > 0. \tag{10}$$

Then, the substitution of formulas (8) and (9) into (7) yields the time derivative of V along the trajectories of the closed-loop system:

$$\dot{V} = -\frac{1}{2}\int_0^\ell \rho h w^2 dx - \left.\frac{\rho g w^2}{2}\right|_{x=\ell} - \left(\varkappa + \gamma k_0 - \frac{\rho(0)g(0)\alpha^2}{2}\right)s^2. \tag{11}$$

4 Stability Analysis

To analyze the stability properties of the above control system, we first perform the change of variables

$$w(x,t) = \tilde{w}(x,t) + \alpha s(t).$$

This allows us to rewrite (4) as a system with zero boundary condition at $x = 0$:

$$\begin{aligned}\frac{\partial \tilde{w}}{\partial t} =& - g\tilde{w}' + v\psi\tilde{w} - \alpha \int_0^\ell \theta(y)\tilde{w}(y,t)dy - \alpha k_1 \tilde{w}(\ell,t)\\ &+ \left(\alpha k_2 + \alpha v\psi - g_c\bar{n}'\right)s - \alpha b u, \quad x \in [0,\ell],\\ \tilde{w}|_{x=0} =& \, 0,\\ \frac{ds}{dt} =& - k_2 s + k_1\tilde{w}|_{x=\ell} + \int_0^\ell \theta(x)\tilde{w}(x,t)dx + bu,\end{aligned} \tag{12}$$

where

$$k_2 = k_0 - \alpha k_1 - \alpha \int_0^\ell \theta(x)dx.$$

Let the function $\rho \in C^1[0,\ell]$ be defined by (10), and let $L^2_\rho(0,\ell)$ denote the weighted L^2-space such that the inner product of $\eta_1, \eta_2 \in L^2_\rho(0,\ell)$ is given by

$$\langle \eta_1, \eta_2 \rangle_{L^2_\rho(0,\ell)} = \int_0^\ell \eta_1(x)\eta_2(x)\rho(x)dx.$$

We also introduce the linear space

$$H = \left\{ \xi = \begin{pmatrix} \eta \\ s \end{pmatrix} \;\middle|\; \eta \in L^2_\rho(0,\ell),\ s \in \mathbb{R} \right\}$$

with the following inner product of elements $\xi_1 = \begin{pmatrix} \eta_1 \\ s_1 \end{pmatrix} \in H$ and $\xi_2 = \begin{pmatrix} \eta_2 \\ s_2 \end{pmatrix} \in H$:

$$\langle \xi_1, \xi_2 \rangle_H = \langle \eta_1 + \alpha s_1, \eta_2 + \alpha s_2 \rangle_{L^2_\rho(0,\ell)} + \gamma s_1 s_2.$$

It is easy to see that H is a Hilbert space if $\gamma > 0$.

Then, system (12) can be represented as the abstract differential equation

$$\frac{d}{dt}\xi(t) = A\xi(t) + Bu, \quad \xi(t) \in H,\ u \in \mathbb{R}, \tag{13}$$

with the unbounded linear operator $A : D(A) \to H$ defined by

$$D(A) = \left\{ \xi = \begin{pmatrix} \eta \\ s \end{pmatrix} \in H \;\middle|\; \eta \in H^1(0,\ell),\ \eta(0) = 0 \right\}, \tag{14}$$

$$\xi = \begin{pmatrix} \eta \\ s \end{pmatrix} \mapsto A\xi = \begin{pmatrix} -g\eta' + v\psi\eta - \alpha \int_0^\ell \theta(y)\eta(y)dy - \alpha k_1\eta(\ell) + s(\alpha k_2 + \alpha v\psi - g_c\bar{n}') \\ -k_2 s + k_1\eta(\ell) + \int_0^\ell \theta(y)\eta(y)dy \end{pmatrix},$$

and

$$B = \begin{pmatrix} -\alpha b \\ b \end{pmatrix} \in H. \tag{15}$$

Here, $H^1(0,\ell)$ denotes the Sobolev space.

The feedback law (8) can be written in the operator form as

$$u = K\xi, \tag{16}$$

where the linear functional $K : D(K) \subset H \to \mathbb{R}$ acts as

$$\xi = \begin{pmatrix} \eta \\ s \end{pmatrix} \mapsto K\xi = -\frac{1}{\gamma b}\left(\varkappa s + \int_0^\ell (\gamma\theta(x) - \rho(x)g_c(x)\bar{n}'(x))\eta(x)dx + \gamma k_1\eta(\ell) \right).$$

We formulate the main stability result for the closed-loop system (13) and (16) as follows.

Theorem 1 *Let the linear operator $\tilde{A} : D(A) \to H$ be defined as $\tilde{A} = A + BK$, where A, B, and K are given by (14), (15), and (16), respectively. Assume, moreover, that the function $\rho \in C^1[0,\ell]$ is defined by (10) with some $h \in C[0,\ell]$ and*

$$\bar{\rho} > 0,\ \gamma > 0,\ \varkappa > \frac{\bar{\rho}g(0)\alpha^2}{2} - \gamma k_0,\ g(\ell) > 0,\ h(x) > 0 \ \text{for all}\ x \in [0,\ell]. \tag{17}$$

Then, the abstract Cauchy problem

$$\begin{aligned} &\frac{d}{dt}\xi(t) = \tilde{A}\xi(t), \quad t \ge 0, \\ &\xi(0) = \xi_0 \in H \end{aligned} \tag{18}$$

is well-posed (in the sense of mild solutions), and the trivial solution of (18) is exponentially stable, i.e.,

$$\|\xi(t)\|_H \le \|\xi_0\|_H e^{-\omega t} \quad \text{for all}\ \xi_0 \in H,\ t \ge 0, \tag{19}$$

with some $\omega > 0$.

Proof A straightforward computation shows that

$$\left\langle \xi, \tilde{A}\xi \right\rangle_H = -\frac{1}{2}\int_0^\ell \rho(x)h(x)\eta^2(x)dx - \frac{\rho(\ell)g(\ell)\eta^2(\ell)}{2} - \left(\varkappa + \gamma k_0 - \frac{\bar{\rho}g(0)\alpha^2}{2}\right)s^2, \tag{20}$$

for all ξ from the dense set $D(\tilde{A}) = D(A) \subset H$. If the conditions (17) hold, then $\left\langle \xi, \tilde{A}\xi \right\rangle_H \le 0$ for all $\xi \in D(\tilde{A})$, which proves that the operator $\tilde{A}$ is dissipative in H. It can also be shown that $\tilde{A}$ is closed, and $A - \lambda I$ is surjective for $\lambda > 0$. Hence, $\tilde{A}$ generates the C_0-semigroup of contractions $\{e^{t\tilde{A}}\}_{t\ge 0}$ on H by the Lumer–Phillips theorem (cf. [23, 24]). The Cauchy problem (18) is thus well-posed on $t \ge 0$, and its mild solutions are defined by

$$\xi(t) = e^{t\tilde{A}}\xi_0, \quad \xi_0 \in H, \ t \ge 0.$$

To prove the exponential decay estimate (19), we analyze the behavior of

$$V(\xi(t)) = \frac{1}{2}\|\xi(t)\|_H^2$$

along the solutions of (18). The above $V(\xi(t))$ plays the same role for the abstract problem (18) as the Lyapunov functional (6) for the closed-loop system (4) and (8).

If $\xi(t)$ is a classical solution of (18) (i.e., $\xi(t) \in D(\tilde{A})$ for all $t \ge 0$), then $\frac{d}{dt}V(\xi(t)) = \left\langle \xi(t), \tilde{A}\xi(t) \right\rangle_H \le 0$. Moreover, the quadratic functional (20) is negative definite with respect to the norm $\|\cdot\|_H$ if the conditions (17) are satisfied, which means that

$$\frac{d}{dt}V(\xi(t)) = \left\langle \xi(t), \tilde{A}\xi(t) \right\rangle_H \le -\frac{\delta}{2}\|\xi(t)\|_H^2 = -\delta V(\xi(t)) \ \text{ for } \ \xi(t) \in D(\tilde{A}) \tag{21}$$

with some constant $\delta > 0$. Then, (19) follows from (21) and the Grönwall–Bellman inequality with $\omega = \delta/2 > 0$. □

5 Preferential Crystallization Model

Consider the 2×2 hyperbolic system with one spatial variable that describes the preferential crystallization of enantiomers [17, 25]:

$$\begin{aligned} &\frac{\partial n_k(x,t)}{\partial t} + G_k(S_k)\frac{\partial n_k(x,t)}{\partial x} = \psi(x)n_k(x,t), \quad x \in [0,\ell], \ t \ge 0, \\ &G_k(S_k)n_k|_{x=0} = B_k(S_k), \quad k = 1, 2, \end{aligned} \tag{22}$$

where $n_1(x,t) \geq 0$ and $n_2(x,t) \geq 0$ are the crystal size distributions for the preferred and counter enantiomers, respectively. Here, $G_k : [1,+\infty) \to \mathbb{R}^+$ characterizes the growth rate of crystals, and $B_k : [1,+\infty) \to \mathbb{R}^+$ describes the nucleation rate of particles of minimum size for the k-th enantiomer. These functions depend on the relative supersaturations $S_1 \geq 1$ and $S_2 \geq 1$ of the preferred and counter enantiomers, which are mutually controlled by using the balance between the incoming and the outgoing mass fluxes in the liquid phase. It is assumed that B_k and G_k are differentiable and strictly increasing functions in their domain of definition such that $B_k(1) = 0$ and $G_k(0) = 0$ for $k = 1,2$. The classification function $\psi(x)$ describing the dissolution of particles below some critical values is assumed to be piecewise continuous on $[0,\ell]$.

It is easy to see that system (22) with $S_k = \bar{S}_k = \text{const} > 1$ has the equilibrium $n_k(x,t) = \bar{n}_k(x)$,

$$\bar{n}_k(x) = \frac{\bar{B}_k}{\bar{G}_k} \exp\left\{\frac{1}{\bar{G}_k}\int_0^x \psi(y)dy\right\}, \quad k = 1,2, \tag{23}$$

where

$$\bar{B}_k = B_k(\bar{S}_k) > 0, \; \bar{G}_k = G_k(\bar{S}_k) > 0.$$

To study the crystallization dynamics in a neighborhood of the steady state (23), we rewrite system (22) with respect to $w_k(x,t) = n_k(x,t) - \bar{n}_k(x)$ as follows:

$$\begin{aligned}
\frac{\partial w_k(x,t)}{\partial t} &= -(\bar{G}_k + \Delta G_k)\frac{\partial w_k(x,t)}{\partial x} + \psi(x)w_k(x,t) - \frac{\Delta G_k}{\bar{G}_k}\psi(x)\bar{n}_k(x), \\
(\bar{G}_k + \Delta G_k)w_k\big|_{x=0} &= \Delta B_k - \Delta G_k\bar{B}_k/\bar{G}_k, \quad k = 1,2,
\end{aligned} \tag{24}$$

where $\Delta B_k = B_k - \bar{B}_k$ and $\Delta G_k = G_k - \bar{G}_k$. Note that the deviations ΔB_k and ΔG_k cannot be controlled independently, as the growth and nucleation rates of both enantiomers mutually depend on mass fractions in the liquid phase. Following the approach of [25], we introduce a scalar variable v that characterizes the deviation of relative saturations from their steady-state values and assume that

$$\begin{aligned}
\Delta G_k/\bar{G}_k &= g_k v + o(|v|), \\
\Delta B_k/\bar{B}_k &= b_k v + o(|v|),
\end{aligned} \tag{25}$$

for small values of v. Thus, the approximation of system (24) takes the form

$$\begin{aligned}
\frac{\partial w_k(x,t)}{\partial t} &= -\bar{G}_k(1 + g_k v)\frac{\partial w_k(x,t)}{\partial x} + \psi(x)w_k(x,t) - g_k\bar{n}_k(x)v, \; x \in (0,\ell), \\
w_k|_{x=0} &= \alpha_k v, \quad k = 1,2,
\end{aligned} \tag{26}$$

where the terms of order $o(|v|)$ are neglected and

$$\alpha_k = (b_k - g_k)\bar{B}_k/\bar{G}_k. \tag{27}$$

We assume further that the rate of change of v can be controlled, i.e.,

$$\frac{dv}{dt} = u, \tag{28}$$

and u is treated as the control.

In control system (26) and (28), the functions $\bar{n}_k(x)$ are defined by (23), and the parameters g_k and α_k are expressed from (25) and (27) in terms of the Taylor coefficients of B_k and G_k.

6 Stabilization with Scalar Input

Similarly to the crystallization model of Sect. 2, we will use weighted L^2-norms to construct a control Lyapunov functional candidate:

$$W(t) = \frac{1}{2}\sum_{k=1}^{2}\int_0^{\ell} \rho_k(x) w_k^2(x,t)\,dx + \frac{\gamma v^2(t)}{2}, \quad \gamma > 0,\ \rho_k(x) > 0. \tag{29}$$

We compute the time derivative of W along the classical solutions of the nonlinear control system (26) and (28) by exploiting the integration by parts and assuming that $w_k(0,t) = \alpha_k v$:

$$\dot{W} = \frac{1}{2}\sum_{k=1}^{2}(W_{0k} + vW_{1k}) + \gamma v u, \tag{30}$$

where

$$W_{0k} = \int_0^{\ell} (\bar{G}_k\rho_k' + 2\rho_k\psi)w_k^2(x,t)\,dx - \bar{G}_k\rho_k(\ell)w_k^2(\ell,t),$$

$$W_{1k} = g_k\int_0^{\ell} (\bar{G}_k w_k\rho_k' - 2\rho_k\bar{n}_k)w_k\,dx + \bar{G}_k(1+g_k v)\rho_k(0)\alpha_k^2 v$$
$$- \bar{G}_k g_k\rho_k(\ell)w_k(\ell,t).$$

To derive a stabilizing control, we choose the density functions $\rho_k(x) > 0$ as solutions to the following differential equations:

$$\bar{G}_k\rho_k'(x) = -2\psi(x)\rho_k(x) - h_k(x)\rho_k(x), \quad x \in [0,\ell],\ k = 1,2. \tag{31}$$

Our main result concerning the stability of the closed-loop system under this above choice of densities $\rho_k(x)$ is summarized below.

Theorem 2 *Let $h_k \in C[0, \ell]$ be such that $h_k(x) \geq h_{k0} > 0$, $k = 1, 2$,*

$$\rho_k(x) = \bar{\rho}_k \exp\left\{-\frac{1}{\bar{G}_k}\int_0^x (2\psi(y) + h_k(y))dy\right\}, \quad \bar{\rho}_k > 0,\ x \in [0, \ell], \tag{32}$$

and let

$$\begin{aligned} u = &-\frac{\varkappa v}{2} + \frac{1}{2\gamma}\sum_{k=1}^{2}\Big\{g_k \int_0^\ell (2\bar{n}_k + (h_k + 2\psi)w_k)w_k\rho_k dx \\ &- \bar{G}_k(1 + g_k v)\bar{\rho}_k\alpha_k^2 v + \bar{G}_k g_k \rho_k(\ell)w_k(\ell, t)\Big\}, \quad \varkappa > 0. \end{aligned} \tag{33}$$

Then, the classical solutions of the closed-loop system (26), (28), *and* (33) *satisfy the following exponential decay estimate:*

$$W(t) \leq W(0)e^{-\omega t}, \quad t \geq 0, \tag{34}$$

where $\omega = \min\{h_{10}, h_{20}, \varkappa\} > 0$.

Proof It is easy to see that the functions $\rho_k(x)$ defined by (32) are general solutions of (31). Then, we transform formula (30) by expressing the control u from (33) and the derivatives of ρ_k from (31). As a result, the time derivative of W along the trajectories of the closed-loop system (26), (28), and (33) reads as follows:

$$\dot{W} = -\frac{1}{2}\sum_{k=1}^{2}\left(\int_0^\ell \rho_k h_k w_k^2 dx + \bar{G}_k\rho_k(\ell)w_k^2(\ell, t)\right) - \frac{\gamma\varkappa}{2}v^2.$$

Then,

$$\dot{W} \leq -\min\{h_{10}, h_{20}, \varkappa\}W,$$

which proves the estimate (34). □

7 Conclusions

The main theoretical contribution of this chapter provides explicit control design schemes for the stabilization of the continuous crystallization model (Theorem 1) and preferential crystallization of enantiomers (Theorem 2). While stability with respect to some integral measure of a population balance model was already analyzed in the paper [18], our results are based on the construction of quadratic

Lyapunov functionals to achieve strong stability in the corresponding L^2-spaces. The efficiency of the proposed controllers remains to be verified by numerical simulations and possible future experimental work.

References

1. Marquardt, W.: Traveling waves in chemical processes. Int. Chem. Eng. **30**, 585–606 (1990)
2. Randolph, A.D., Larson, M.A.: Theory of Particulate Processes. Academis, New York (1988)
3. Ramkrishna, D.: Population balances. In: Theory and Applications to Particulate Systems in Engineering. Academis, San Diego (2000)
4. Grüner, S., Kienle, A.: Equilibrium theory and nonlinear waves for reactive distillation columns and chromatographic reactors. Chem. Eng. Sci. **59**, 901–918 (2004)
5. Guiochon, G., Felinger, A., Shirazi, D.G.: Fundamentals of Preparative and Nonlinear Chromatography, 2nd edn. Academic, Amsterdam (2006)
6. Bastin, G., Coron J.-M.: Stability and Boundary Stabilization of 1-D Hyperbolic Systems. Springer, Cham (2016)
7. Kokotovic, P.: The joy of feedback: nonlinear and adaptive. IEEE Control Syst. Mag. **12**(3), 7–17 (1992)
8. Krstic, M., Smyshlyaev, A.: Boundary Control of PDEs: A Course on Backstepping Designs. SIAM, Philadelphia (2008)
9. Lamare, P.-O., Bekiaris-Liberis, N.: Control of 2×2 linear hyperbolic systems: backstepping-based trajectory generation and PI-based tracking. Syst. Control Lett. **86**, 24–33 (2015)
10. Deutscher, J., Kerschbaum, S.: Output regulation for coupled linear parabolic PIDEs. Automatica **100**, 360–370 (2019)
11. Fliess, M., Lévine, J., Martin, Ph., Rouchon, P.: Flatness and defect of nonlinear systems: introductory theory and examples. Int. J. Control **61**, 1327–1361 (1995)
12. Rothfuss, R., Rudolph, J., Zeitz, M.: Flatness based control of a nonlinear chemical reactor model. Automatica **32**, 1433–1439 (1996)
13. Meurer, T.: Flatness of a class of linear Volterra partial integro-differential equations. IFAC-PapersOnLine **49**, 174–179 (2016)
14. Zuyev, A., Benner, P.: Local steering problem for a class of control-affine systems with application to continuous crystallization processes. PAMM **16**, 831–832 (2016). https://doi.org/10.1002/pamm.201610404
15. Zuyev, A., Grushkovskaya, V.: Motion planning for control-affine systems satisfying low-order controllability conditions. Int. J. Control **90**, 2517–2537 (2017)
16. Vollmer, U., Raisch, J: H_∞-Control of a continuous crystallizer. Control Eng. Practice **9**, 837–845 (2001)
17. Qamar, S., Ashfaq, A., Angelov, I., Elsner, M.P., Warnecke, G., Seidel-Morgenstern, A.: Numerical solutions of population balance models in preferential crystallization. Chem. Eng. Sci. **63**(5), 1342–1352 (2008)
18. Palis, S., Kienle, A.: Discrepancy based control of particulate processes. J. Process Control **24**(3), 33–46 (2014)
19. Martynyuk, A.M., Slyn'ko, V.I.: On stability of linear hybrid mechanical systems with distributed components. Ukrainian Math. J. **60**, 235–252 (2008)
20. Zuiev, A.L.: On Brockett's condition for smooth stabilization with respect to a part of the variables. 1999 European Control Conference (ECC), 1729-1732 (1999). https://doi.org/10.23919/ECC.1999.7099564
21. Zuyev, A.L.: Stabilization of non-autonomous systems with respect to a part of variables by means of control Lyapunov functions. J. Autom. Inf. Sci. **32**, 18–25 (2000). https://doi.org/10.1615/JAutomatInfScien.v32.i10.30

22. Zuyev, A., Kienle, A., Benner, P.: Construction of a Lyapunov functional for a class of controlled population balance models. PAMM **17**, 827–828 (2017). https://doi.org/10.1002/pamm.201710381
23. Pazy, A.: Semigroups of Linear Operators and Applications to Partial Differential Equations. Springer, New York (1983)
24. Barbu, V..: Nonlinear Differential Equations of Monotone Types in Banach Spaces. Springer, New York (2010)
25. Zuyev, A., Benner, P.: Control design for a 2 × 2 hyperbolic system with application to preferential crystallization. PAMM **18** (2018). https://doi.org/10.1002/pamm.201800455

The manufacturer's authorised representative in the EU is Springer Nature Customer Service Centre GmbH, Europaplatz 3, 69115 Heidelberg, Germany. If you have any concerns regarding our products, please contact ProductSafety@springernature.com

Printed and bound by CPI Group (UK) Ltd, Croydon, CR0 4YY
30/09/2026
02225467-0015